2nd Louhi Workshop on Text and Data Mining of Health Documents 2010

(Louhi-10)

Held at NAACL HLT 2010

Los Angeles, California, USA
5 June 2010

ISBN: 978-1-61738-680-0

Printed from e-media with permission by:

Curran Associates, Inc.
57 Morehouse Lane
Red Hook, NY 12571

Some format issues inherent in the e-media version may also appear in this print version.

Printed by Curran Associates, Inc. (2010)

Association for Computational Linguistics
209 N. Eighth Street
Stroudsburg, Pennsylvania 18360

Phone: 1-570-476-8006
Fax: 1-570-476-0860

acl@aclweb.org

Additional copies of this publication are available from:

Curran Associates, Inc.
57 Morehouse Lane
Red Hook, NY 12571 USA
Phone: 845-758-0400
Fax: 845-758-2634
Email: curran@proceedings.com
Web: www.proceedings.com

Preface

Welcome to the Second Louhi Workshop on Text and Data Mining of Health Documents, Louhi 2010, in Los Angeles, California, USA. We aim to bring together researchers and practitioners in a multidisciplinary conference on new uses of computer systems for data mining in the health care area. The very first Louhi Conference 2008 in Turku, Finland, showed the diversity and complexity of these issues. The increasing access to clinical data from health care systems and the progress of new methods and approaches in text and data mining results in a growing attention from the perspective of patient care as well as biomedical research and education. Improving performance in medical information retrieval is a challenge of complex nature requiring several approaches. Automatic indexing of clinical findings, observations, diseases and treatments is of high relevance for clinical work and research in general. Implementing decision support and guidelines to reach evidence-based practice is a specific area of interest (Fiszman et al. 2000). Detecting adverse event is another field that needs a high-quality system to detect findings and events in clinical documents (Griffin and Resar 2009). In several other areas indexing and mining in clinical text are of utmost importance – from everyday clinical work to translational biomedical research (Meystre et al. 2008). Health care consumer web sites as well as news web sites contain important information worthwhile monitoring to extract both information on specific diseases directed to the layman as well as epidemiological information. The papers presented in this workshop aim at exploring computational methods and tools to improve and support the work in these different fields. The Second Louhi workshop will continue to focus and reflect on computer use in every-day clinical work in hospitals and clinics such as electronic record systems, computer aided summaries, clinical coding, computerized clinical guidelines, computer decision systems, as well as related ethical concerns and security. Much of this work concerns itself by necessity with computer aided language use, and as such Louhi aims at providing an arena for report on development in a diversity of languages. In the papers presented at Louhi 2010 we can read about many of the challenges identified above.

A short description of each paper follows in order of appearance. In the first paper, *Friberg Heppin* describes the Swedish MedEval corpus, which has been annotated for the needs of both physicians and patients. The corpus has then been indexed with two different methods for an information retrieval experiment that aims to satisfy the requirements of both user groups. *Bhatia et al.* extracted information from English electronic health records containing diabetes information from a large number of patients, with the aim to detect populations at high risk of diabetes. *Skeppstedt* has ported the negation detection system NegEx, which is written for English clinical text, to Swedish, and describes the porting process in detail and finally evaluates the Swedish version of NegEx. *Schreitter et al.* describe a system for automatic speech recognition of dictated medical records in English. The aim of their work was to reduce errors in recognizing medication names, trademarks, dosages and strengths, and the authors use the Unified Medical Language System (UMLS) as a knowledge base for the recognition. *von Etter et al.* present an approach on monitoring epidemic information from online news articles, with epidemic intelligence officers as the intended user group. They have defined guidelines based on correctness and reliability together with the medical users and further annotated 1 000 articles that were then utilized in a machine learning based classification experiment. The paper by *Martin* is closely related to the work by *von Etter et al.* described above, but focuses on English web pages containing health care information directed to health care consumers. The author describes an annotation scheme based on type of information, applied by two students annotating 200 pages of health information documents.

Roque et al. present an overview of five open source visualization tools for electronic health records for medical practitioners. They describe the tools in the context of users, goals and tasks focusing on the temporal aspects of the visual presentation. *Melton et al.* present a system for identifying the long form of acronyms and abbreviations in biomedical text, using MetaMap applied on the UMLS to identify the long forms by expanding the acronyms. In *Allvin et al.* the authors have carried out both a qualitative and quantitative comparative study of Finnish and Swedish nursing narratives from two intensive care units. As the Swedish and Finnish languages belong to different language groups, while the countries are culturally closely related, this study explores how this might influence what is expressed in the narratives. *Halgrim et al.* describe a hybrid system for medical extraction based on both rule based and statistical classifiers. The system is applied on English narrative clinical records from the i2b2 challenge and uses several rule based processing steps where field detection is one significant step detecting if a medication occurs in narrative text or in a list of medications. *Kokkinakis and Toporowska Gronostaj* have carried out a pilot study to extract events from Swedish medical and clinical text, based on Frame Semantics and their methodology Swedish FrameNet++ (SFN++). *Hirschman and Aberdeen* have developed new metrics for the de-identification and the re-identification problem of clinical text. They emphasize that traditional information extraction metrics are not enough to address the real-world questions on "how good are current de-identification systems?". *Medori and Fairon* present a semi-automatic system for assigning ICD-9-CM codes to discharge summaries in French, and shows that stemmed and extracted specific encoding information gives better classification results than without pre-processing. Finally, *Lin et al.* present ongoing work using Conditional Random Fields to extract important information from clinical research articles, focusing on extracting formulaic information, metadata about the authors, longitudinal data and medical intervention methods.

We received in total 16 submissions from eleven countries and three continents, and after a rigorous double-blind peer-review process we could accept 14 of these submissions to be published in the Louhi 2010 proceedings.

Dear reader, most welcome to study this proceeding, which we hope will raise interest and open new perspectives in text and data mining of health documents.

Stockholm, April 2010

Hercules Dalianis, Martin Hassel and Gunnar Nilsson

References

Marcelo Fiszman and Peter J. Haug. 2000. Using medical language processing to support real-time evaluation of pneumonia guidelines. *Proceedings of AMIA Annual Symposium 2000*; 235-9.

Frances A. Griffin and Roger K.Resar. 2009. IHI Global Trigger Tool for Measuring Adverse Events (Second Edition). *IHI Innovation Series white paper.* Cambridge, MA: Institute for Healthcare Improvement; 2009.

Stéphane M. Meystre, Guergana K. Savova, Karin C. Kipper-Schuler, and. John E. Hurdle. 2008. Extracting information from textual documents in the electronic health record: a review of recent research. In: *IMIA Year-book of Medical Informatics 2008.* Methods Inf Med 2008; 47 Suppl 1:138-154.

Chair:

Hercules Dalianis, DSV/Stockholm University

Program co-chairs:

Martin Hassel, DSV/Stockholm University
Gunnar Nilsson, Karolinska Institutet

Organizers:

Hercules Dalianis, email: hercules@dsv.su.se
Martin Hassel, email: xmartin@dsv.su.se
Sumithra Velupillai, email: sumithra@dsv.su.se
Address: DSV, Stockholm University, Forum 100, 164 40 Kista, Sweden

Program Committee:

Sophia Ananiadou, University of Manchester, UK
Stephen Anthony, University of New South Wales, Australia
Henrik Boström, Stockholm University
Søren Brunak, Technical University of Denmark, DTU
Wendy Chapman, University of Pittsburgh
Aaron Cohen, Oregon Health & Science University
Richárd Farkas, University of Szeged, Hungary
Filip Ginter, University of Turku, Finland
Helena Karsten, Åbo Akademi, Finland
Dimitrios Kokkinakis, University of Gothenburg
Anette Hulth, Swedish Institute for Infectious Disease Control, Sweden
Sabine Koch, Karolinska institutet, Sweden
Jong C. Park, KAIST, South Korea
Tapio Pahikkala, University of Turku, Finland
Serguei Pakhomov, Center for Clinical and Cognitive Neuropharmacology, University of Minnesota, USA
Jon D. Patrick, University of Sydney, Australia
Sampo Pyysalo, University of Tokyo
Tapio Salakoski, University of Turku, Finland
Sanna Salanterä, University of Turku, Finland
Laura Slaughter, NTNU, Norway
Hanna Suominen, University of Turku, Finland
György Szarvas, UKP Lab, Technical University of Darmstadt, Germany
Özlem Uzuner, University at Albany, State University of New York, USA
Jaak Vilo, University of Tartu, Estonia
Pierre Zweigenbaum, LIMSI, France
Hans Åhlfeldt, Linköping University, Sweden

Publication chair:

Hercules Dalianis, DSV/ Stockholm University

Local organizing chair:

Sumithra Velupillai, DSV/ Stockholm University

Invited Speaker:

Eduard Hovy, Information Sciences Institute of the University of Southern California

Table of Contents

Workshop Program

Saturday, June 5, 2010

Session I

8:45–9:00 Opening Remarks

9:00–10:00 Invited Talk: Creating Training Material for Health Informatics: Toward a Science of Annotation, Eduard Hovy

10:00–10:30 *MedEval- A Swedish Medical Test Collection with Doctors and Patients User Groups*
Karin Friberg Heppin

10:30–11:00 **Morning break**

Session II: Paper presentations

11:00–11:30 *Extracting Information for Generating A Diabetes Report Card from Free Text in Physicians Notes*
Ramanjot Singh Bhatia, Amber Graystone, Ross A Davies, Susan McClinton, Jason Morin and Richard F Davies

11:30–12:00 *Negation Detection in Swedish Clinical Text*
Maria Skeppstedt

12:00–12:30 *Using Domain Knowledge about Medications to Correct Recognition Errors in Medical Report Creation*
Stephanie Schreitter, Alexandra Klein, Johannes Matiasek and Harald Trost

12:30–2:00 **Lunch break**

Saturday, June 5, 2010 (continued)

Session III: Paper presentations

2:00–2:30 *Assessment of Utility in Web Mining for the Domain of Public Health*
Peter von Etter, Silja Huttunen, Arto Vihavainen, Matti Vuorinen and Roman Yangarber

2:30–3:00 *Reliability and Type of Consumer Health Documents on the World Wide Web: an Annotation Study*
Melanie Martin

3:00–3:30 **Afternoon break**

Session IV: Poster presentations

3:00–4:00 *Automated Identification of Synonyms in Biomedical Acronym Sense Inventories*
Genevieve B. Melton, SungRim Moon, Bridget McInnes and Serguei Pakhomov

Characteristics and Analysis of Finnish and Swedish Clinical Intensive Care Nursing Narratives
Helen Allvin, Elin Carlsson, Hercules Dalianis, Riitta Danielsson-Ojala, Vidas Daudaravicius, Martin Hassel, Dimitrios Kokkinakis, Heljä Lundgren-Laine, Gunnar Nilsson, Øystein Nytrø, Sanna Salanterä, Maria Skeppstedt, Hanna Suominen and Sumithra Velupillai

Extracting Medication Information from Discharge Summaries
Scott Halgrim, Fei Xia, Imre Solti, Eithon Cadag and Özlem Uzuner

Linking SweFN++ with Medical Resources, towards a MedFrameNet for Swedish
Dimitrios Kokkinakis and Maria Toporowska Gronostaj

Measuring Risk and Information Preservation: Toward New Metrics for De-identification of Clinical Texts
Lynette Hirschman and John Aberdeen

A Comparison of Several Key Information Visualization Systems for Secondary Use of Electronic Health Record Content
Francisco Roque, Laura Slaughter and Aleksandr Tkatšenko

Saturday, June 5, 2010 (continued)

MedEval — A Swedish Medical Test Collection with Doctors and Patients User Groups

Karin Friberg Heppin
Department of Swedish
University of Gothenburg
Gothenburg, Sweden
karin.friberg@svenska.gu.se

Abstract

MedEval is a Swedish medical test collection where assessments have been made, not only for topical relevance, but also for target reader group: Doctors or Patients. The user of the test collection can choose if s/he wishes to search in the Doctors or the Patients scenarios where the topical relevance assessments have been adjusted with consideration to user group, or to search in a scenario which regards only topical relevance. MedEval makes it possible to compare the effectiveness of search terms when it comes to retrieving documents aimed at the different user groups. MedEval is also the first medical Swedish test collection.

1 A New Test Collection

When the decision was made to build a new test collection, the Department of Swedish at the University of Gothenburg was involved in projects of research in medical language processing. There was also a growing interest of research in information retrieval. There existed no Swedish medical test collection. Creating one seemed to be a good investment in knowledge and resources, even though this involved a team of people during many months. As building a test collection is a major undertaking not many exist. OHSUMED is a medical test collection, albeit in English. It is built on nearly 350 000 references from MEDLINE. The OHSUMED documents are assessed on a three graded scale: definitely, possibly and not relevant. OHSUMED contains 106 topics generated by physicians from authentic situations. The topics consist of both information about the patient and the request. (OHSUMED, 2007)

With a new collection such as MedEval, the Swedish department could take control over the architecture and make decisions such as using a four graded scale of relevance, making it possible to employ a variety of evaluation tools. However, the most important decision was to assess documents, not only for relevance to topics, but also for intended groups of readers, 'Doctors: medical professionals' or 'Patients: lay persons', and to allow the user to choose user scenario: None, Doctors or Patients.

2 Documents

The MedEval test collection is built on documents from the MedLex medical corpus (Kokkinakis, 2004). MedLex consists of scientific articles from medical journals, teaching material, guidelines, patient FAQs, health care information, etc. The set of documents used in MedEval is a snapshot of MedLex in October 2007, approximately 42 200 documents or 15 million tokens (see table 1). The documents are stored in the trectext format.

3 Indexes

The MedEval test collection has two indexes. One where the documents are converted to lower case, tokenized and lemmatized, and one where the compounds also are decomposed. In the second index, the compound terms are indexed as a whole together with the compound constituents. For instance: the compound *saltkoncentration* 'salt concentration' is indexed as *saltkoncentration*, *salt*, and *koncentration*.

Proceedings of the NAACL HLT 2010 Second Louhi Workshop on Text and Data Mining of Health Documents, pages 1–7,
Los Angeles, California, June 2010.

Type of source	Number of documents	Percent of documents	Number of tokens	Percent of tokens
Journals and periodicals	8 453	20.0	5.3 million	34.6
Specialized sites	14 631	34.6	2.9 million	19.1
Pharmaceutical companies	9 200	21.8	2.3 million	14.8
Government, faculties, institutes, and hospitals	2 955	7.0	2.0 million	13.3
Health-care communication companies	4 036	9.6	1.7 million	11.3
Media (TV, daily newspapers)	2 980	7.1	1.0 million	6.9
Total	42 255	100.1	15.2 million	100

Table 1: The genres of the MedEval document sources. The document collection is a snapshot of the MedLex corpus in October 2007. (D. Kokkinakis, p.c.)

4 Topics

Two medical students in their fourth year of studies were hired to create the topics. Their instructions were to create information needs that could be requested in real medical situations. 100 topics were created in the first stage. 62 of these were used in the collection.

A topic consists of a title, a description and a narrative. The title is a short phrase summarizing the information need. The description is concise information about the topic, usually in the form of a question or a request. The narrative is a few sentences long and it stipulates what makes a document relevant to the topic. The narrative contains the guidelines for the assessors when judging the relevance of the documents in the next stage. An example of a topic is given below. The English equivalent of the description of topic 51 is: *Why can a patient with cancer contract anemia?*

<TOP>
<TOPNO>51</TOPNO>
<TITLE> Anemi och cancer </TITLE>
<DESC> Varför kan en patient med cancer drabbas av anemi? </DESC>
<NARR> Relevanta dokument ska innehålla information om vad anemi /blodbrist är, symtom, behandling och orsaker. Information om cancerrelaterad anemi dels utlöst av cancern och dels utlöst av cancerbehandlingen är relevant. </NARR>
</TOP>

5 Selecting Documents to Assess

In the ideal test collection every document would be assessed for relevance with respect to every topic. But with over 42 000 documents and 62 topics, taking 8 minutes to assess each document, it would take four persons more than 40 years working 40 hours per week to finish the assessments.

Instead, only the documents that were considered most likely to be relevant to each topic were assessed. The documents were filtered out by use of four queries, one specific and one exhaustive for each index. The documents selected for each topic were sorted by document ID and duplicates were removed. This was done so that the assessors would not know how high a document had been ranked, or in how many searches it had been retrieved. For each topic and each of the four queries the 100 highest ranked documents were selected, if, in fact, there were that many.

6 Relevance Judgments

For the relevance judgments four new medical students were consulted. For each of 62 topics, an assessor read through the documents to be assessed and decided, for each document, the intended group of readers and the degree of relevance to the topic. The documents for each individual need were assessed by one and the same assessor for reasons of consistency.

The MedEval relevance assessments were made on a four graded scale, 0-3, where 0 is ‘Not at all relevant’ and 3 is ‘Highly relevant’. The scale is easily turned into a binary scale by stating that the documents with the lower grades are to be considered non-relevant and the ones with higher grades relevant. Where the division is made between relevant and non-relevant depends on the needs of the user in each case.

The relevance considered by the assessors was topical relevance, how well a document corresponds to a topic. The assessors were instructed not to involve user relevance in this score. Each document was judged on its own merits. The novelty of the contents of a document should not be considered.

7 Target Groups

In addition to topical relevance the assessors judged each document for reader target group, that is which group of readers was the intended: Patients, if a document was written for lay persons, or Doctors, if it was written for medical professionals.

For a classification of documents according to intended reader group to be useful, there must be a measureable difference between the document classes. Table 2 shows a number of type/token frequencies in different subsets of the collection. In each set duplicates were removed in the case that a document had been assessed for more than one topic. The subsets considered are described below. Full form types are the original terms of the documents before lemmatization and lemma types are the same terms after lemmatization.

Entire collection All documents of the MedEval collection.

Assessed documents All documents that have been assessed for any topic.

Doctors assessed All documents that for at least one topic have been assessed to have target group Doctors.

Patients assessed All documents that for at least one topic have been assessed to have target group Patients.

Common files All documents that for at least one topic have been assessed to have target group Doctors and for another to have target group Patients.

Doctors relevant All documents that for at least one topic have been assessed to have at least relevance grade 1 and to have target group Doctors.

Patients relevant All documents that for at least one topic have been assessed to have at least relevance grade 1 and to have target group Patients.

Before counting frequencies, the files were cleaned from tags, IDs, dates (in the date tag, not in the actual text), web information and punctuation marks. Some observations are readily made by studying table 2.

The number of tokens per document is significantly smaller for the entire collection, than for any subset. This means that there is a large number of short documents that were not retrieved by any query when the documents to be assessed were selected. Maybe not surprising, since short documents contain few terms which can match the queries.

The documents in the set 'Patients assessed' had only 57% the number of tokens per document, compared to the documents in 'Doctors assessed'. Even though there were over 1 000 more documents in 'Patients assessed' than in 'Doctors assessed', there were over 50 000 more lemma types in the doctor documents and almost 30 000 more lemma compound types. The average word length in 'Doctors assessed' was 6.29 compared to 5.73 for 'Patients assessed'. The ratio of compound tokens was also higher in the doctor documents, 0.128 compared to 0.098.

Table 3 illustrates the fact that the doctor documents contain more and longer terms and more compounds than patient documents. This table shows frequencies of all full form types of strings beginning with *förmak* 'atria' in 'Patients assessed' and 'Doctors assessed' respectively. The patient documents have 18 full form types beginning with *förmak* while doctor documents have 75. That is more than four times more types for the doctor documents.

A closer look at the frequencies of *förmak*∗ in the professional and lay person texts reveals that not all frequencies are higher for professionals. The frequencies of nouns in the definite form in the lay person texts are close to, equal or higher than the same forms in the professional texts.

	Entire collection	Assessed documents	Doctors assessed	Patients assessed	Common files	Doctors relevant	Patients relevant
Number of documents	42 250	7 044	3 272	4 334	562	1 233	1 654
Tokens	12 991 157	5 034 323	3 232 772	2 431 160	629 609	1 361 700	988 236
Tokens/document	307	715	988	561	1 120	1 104	596
Average word length	5.75	6.04	6.29	5.73	6.16	6.33	5.63
Full form types	334 559	181 354	154 901	92 803	50 961	87 814	43 825
Lemma types	267 892	146 631	126 217	73 121	40 857	71 974	34 263
Compound tokens	1 273 874	573 625	412 475	237 267	76 117	179 580	92 420
Full form compound types	187 904	99 614	83 846	47 387	24 083	45 257	20 157
Lemma compound types	144 159	78 508	66 907	37 151	19 685	36 867	16 006
Ratio of compounds	0.098	0.114	0.128	0.098	0.120	0.132	0.094

Table 2: Type and token frequencies of the terms in different subsets of the MedEval test collection.

Looking at all instances of strings beginning with *förmak*∗ in the two sets of documents there is a significant difference. In the patient documents 66 tokens of 372, or 17.7%, are nouns in the definite form, while the corresponding numbers for the doctor documents is 89 of 932 tokens, or 9.6%. At this stage one can only speculate why this is so. A hypothesis is that doctors/medical professionals often discuss matters in a generic point of view, while patients/lay persons discuss specific cases.

Term	Doctors	Patients
förmaken	21	21
förmakens	1	2
förmaket	11	14
förmaksflimret	16	28
förmaksmyocyterna	2	1

Table 4: Frequencies of terms beginning with *förmak* 'atria', which are in the definite form in the set 'Patients assessed'. The frequencies of these word forms in the documents written for the two target groups are compared.

8 User Groups

The MedEval test collection allows the user to state user group: *None* (no specified group), *Doctors* or *Patients*. This choice directs the user to one of three scenarios. The None scenario contains the topical relevance grades as made by the assessors. The Doctors scenario contains the same grades with the exception that the grades of the documents marked for Patients target group are downgraded by one. In the same way the Patients scenario has the documents marked for Doctors target group downgraded by one. This means that for a doctor user patient documents originally given relevance 3, are graded with 2, documents given relevance 2 are graded 1 and documents given relevance 1 are graded 0. The same is done in the Patients scenario with the doctor documents. The idea is that a document that is written for a reader from one target group but retrieved for a user from the other group will not be non-relevant, but less useful than a document from the correct target group. Put differently, a document intended for patients would contain information that doctors (hopefully) already know. On the other hand, documents intended for doctors, even though they might be topically relevant for a patient's need, run a great risk of being written in such a way that a patient will have problems grasping the whole content.

Adjusting relevance in the manner described affects the scenario recall bases. Since relevance grades are downgraded for documents of the opposing target group there will be fewer relevant documents in the Doctors and Patients scenarios than in the None scenario. This is demonstrated in figure 1 where the ideal cumulated gain for the three scenarios of topics 28, 36 and 92 are shown. The ideal cumulated gain is the maximum score of retrieved information possible at each position in a ranked list of documents (Järvelin, Kekäläinen, 2002). The score for each position is the sum of all relevance scores so far in the ranked list.

The three topics of figure 1 show different characteristics with reference to the number of relevant

Lay person audience	förmak	73	förmaksflimmer	219
	förmaken	21	förmaksflimmerattacker	1
	förmakens	2	förmaksflimmerpatienter	1
	förmaket	14	förmaksflimret	28
	förmaks	1	förmakslimmer	1
	förmaksarytmier	2	förmaksmyocyterna	1
	förmakseffekt	1	förmakstakykardi	1
	förmaksfladder	2	förmaksutlösta	2
	förmaksflimer	1	förmaksöra	1
Professional audience	förmak	93	förmaksmuskelns	1
	förmaken	21	förmaksmuskulaturen	2
	förmakens	1	förmaksmyocyterna	2
	förmaket	11	förmaksmyokard	3
	förmakets	1	förmaksmyokardiet	1
	förmaks	21	förmaksmyxom	2
	förmaksaktivering	1	förmaksnivå	2
	förmaksaktivitet	1	förmaksnära	1
	förmaksaktiviteten	2	förmaksoch	1
	förmaksanatomi	1	förmakspacing	7
	förmaksarytmi	2	förmakspeptider	1
	förmaksarytmier	9	förmaksrytmer	1
	förmaksbidraget	1	förmaksseptostomi	1
	förmaksbradyarytmi	1	förmaksseptum	2
	förmaksdefibrillator	2	förmaksseptumaneurysm	10
	förmakseffekt	2	förmaksseptumdefekt	5
	förmaksfladder	57	förmaksseptumdefekten	1
	förmaksfladdret	2	förmaksseptumdefekter	1
	förmaksflimmer	544	förmaksseptums	1
	förmaksflimmerablationer	2	förmaksstimulerat	1
	förmaksflimmerattacker	1	förmaksstimulerin	5
	förmaksflimmerduration	2	förmaksstorlek	2
	förmaksflimmerepisoder	4	förmaksstorleken	1
	förmaksflimmerfladder	2	förmakssynkron	1
	förmaksflimmerpatienter	4	förmakssystole	1
	förmaksflimmerrecidiv	1	förmakstaket	1
	förmaksflimmertendensen	1	förmakstakykardi	11
	förmaksflimmerunderhållande	1	förmakstakykardie	8
	förmaksflimret	16	förmakstromb	2
	förmaksflimrets	4	förmakstryck	1
	förmaksfrekvenser	1	förmakstrycket	1
	förmaksfunktion	1	förmaksvolym	2
	förmaksförstoring	1	förmaksvägg	1
	förmaksimpuls	1	förmaksväggarna	2
	förmaksinhiberad	1	förmaksväggen	6
	förmakskontraktion	4	förmaksvävnaden	2
	förmakskontraktionen	6	förmaksöra	9
	förmakskontraktionens	1	förmaksöronen	2
	förmaksmuskeln	1		

Table 3: This is a randomly chosen example of the difference in the number of types and of tokens in the documents written for a lay person audience, in the set 'Patients assessed' and the ones written for a professional audience, in the set 'Doctors assessed'. The table shows all types of strings beginning with *förmak* 'atria' in documents written for the two target groups. The number of tokens for each type is also shown.

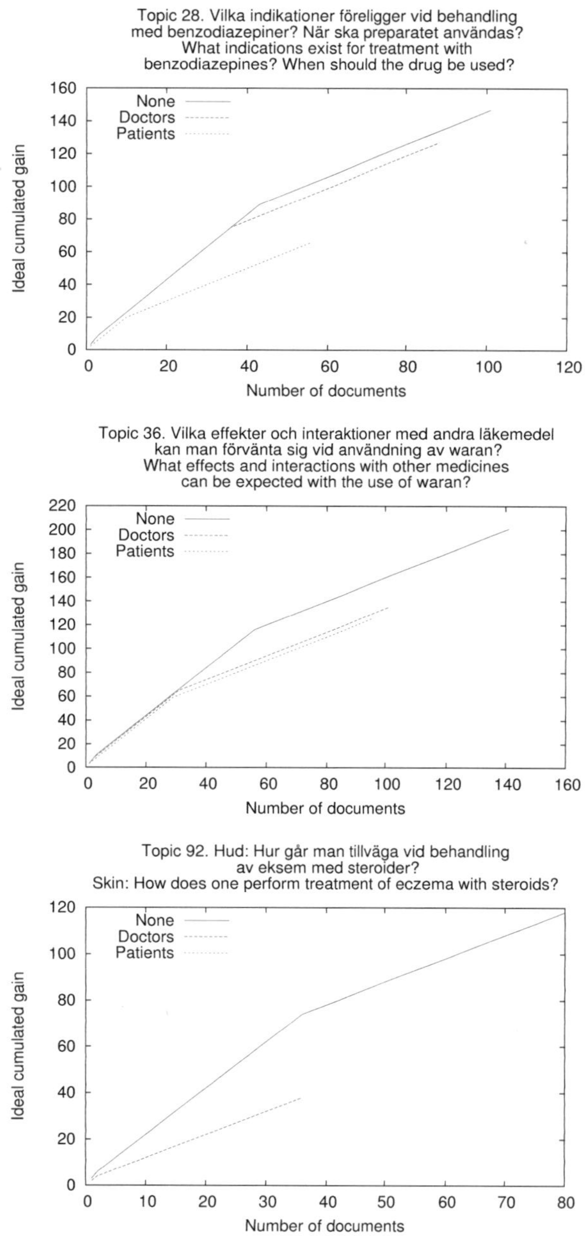

Figure 1: The recall bases of topic 28, 36 and 92 represented in ideal cumulated gain for the three scenarios: None, Doctors and Patients. For topic 28 most of the highly relevant and fairly relevant documents were assessed to have target group Doctors. Topic 36 had the relevant documents spread fairly evenly between the Doctors and Patients target groups. Topic 92 showed no documents of any relevance grade for documents marked for target group Doctors. Thus the None and the Patients ideal gain vector coincide fully, while the cumulated gain for the Doctors scenario is very low.

doctor and patient documents. Topic 36 has fairly similar cumulated gain curves for the Doctors and Patients scenarios. Topic 28 has a majority of doctor documents, while topic 92 had no documents of any relevance grade for documents marked for target group Doctors. Thus the None and the Patients ideal gain vector coincide fully, while the cumulated gain for the Doctors scenario is very low, originating from downgraded patient documents.

9 Example Runs

To demonstrate the effectiveness of search terms from the different styles of language of the two target groups, the synonyms *anemi* 'anemia' and *blodbrist* 'blood lack' were run as search keys for topic 51 in the Doctors and Patients scenarios. *anemi* is a neoclassical term, belonging to the professional language and *blodbrist* is the corresponding lay person term.

In the Doctors scenario the difference between the results of the two search keys was striking: full recall for the neoclassical term quite early in the ranked list of documents and no recall at all for the lay person term. The Patients scenario did not show as big difference between the search keys. Note that the resulting ranked lists of documents is the same for both scenarios for the same search key. It is the relevance grades of the retrieved documents that differ.

Scenario	Recall	anemi	blodbrist
Doctors	@10	50% (4/8)	0% (0/8)
	@20	100% (8/8)	0% (0/8)
	@100	100% (8/8)	0% (0/8)
Patients	@10	22% (4/18)	33% (6/18)
	@20	39% (7/18)	39% (7/18)
	@100	66% (12/18)	56% (10/18)

Table 5: Running the synonyms *anemi* 'anemia' and *blodbrist* 'blood lack' as search keys for topic 51 in the Doctors scenario gave full recall early in the ranking list for the neoclassical term *anemi*, but no recall at all for the lay person term *blodbrist*. In the Patients scenario the difference in effectiveness for these search keys was not as striking.

10 Final Words

This paper shows a few aspects of medical information retrieval which can be studied with the use of the MedEval test collection. The main novelty of the collection is the marking of document target groups, Doctors and Patients, together with with the possibility to choose user group. This opens up new areas of research in Swedish information retrieval such as how one can retrieve documents suited for different groups of users.

The Department of Swedish at the University of Gothenburg is in the process of making the MedEval test collection available to academic researchers.

Acknowledgments

The author would like to thank the FIRE (Finnish Information Retrieval Experts) research group at the University of Tampere, Finland, for their invaluable help in building the MedEval test collection.

References

OSHUMED. 2007. *The OHSUMED test collection.* [www] <http://ir.ohsu.edu/ohsumed/ohsumed.html>.

Kalervo Järvelin and Jaana Kekäläinen. 2002. Cumulated gain-based evaluation of IR techniques. *ACM Transactions on Information Systems*, Vol. 20, No.4, pages 422-446.

Dimitrios Kokkinakis. 2004. *Medlex: Technical report.* Department of Swedish, University of Gothenburg, Sweden. [www] <http://demo.spraakdata.gu.se/svedk/pbl/MEDLEX_work2004.pdf>.

Extracting Information for Generating A Diabetes Report Card from Free Text in Physicians Notes

Ramanjot S Bhatia
University of Ottawa Heart Institute
Ottawa, Ontario.
Rbhatia@ottawaheart.ca

Amber Graystone
McMaster University
Hamilton, Ontario.
amber.graystone@medportal.ca

Ross A Davies
University of Ottawa Heart Institute
Ottawa, Ontario.
RADavies@ottawaheart.ca

Susan McClinton
University of Ottawa Heart Institute
Ottawa, Ontario.
SMcClinton@ottawaheart.ca

Jason Morin
National Research Council Canada
Ottawa, Ontario.
jason.morin@nrc-cnrc.gc.ca

Richard F Davies
University of Ottawa Heart Institute
Ottawa, Ontario.
RFDavies@ottawaheart.ca

Abstract

Achieving guideline-based targets in patients with diabetes is crucial for improving clinical outcomes and preventing long-term complications. Using electronic heath records (EHRs) to identify high-risk patients for further intervention by screening large populations is limited because many EHRs store clinical information as dictated and transcribed free text notes that are not amenable to statistical analysis. This paper presents the process of extracting elements needed for generating a diabetes report card from free text notes written in English. Numerical measurements, representing lab values and physical examinations results are extracted from free text documents and then stored in a structured database. Extracting diagnosis information and medication lists are work in progress. The complete dataset for this project is comprised of 81,932 documents from 30,459 patients collected over a period of 5 years. The patient population is considered high risk for diabetes as they have existing cardiovascular complications. Experimental results validate our method, demonstrating high precision (88.8-100%).

1 Introduction

A standard practice for care providers is to record patient consults using voice dictation. The voice dictation record is transcribed into free text and stored electronically. The nature of this text is narrative with a possibility of containing headings marking the boundaries of the paragraphs. This remains the medium of choice for storing key patient information as opposed to structured tables due to time constraints, uncertainty about the use of codes, classification limitations, and difficulty with the use of computer systems. The information being stored in machine readable format is not amenable to any form of statistical analysis or review as it exists (Mcdonald 1997, Lovis et al. 2000). The usefulness of mining information from this text has been stressed by many including Heinze et al. (2001) and Hripcsak et al. (1995). The information unlocked from the free text could be used for facilitating patient management, researching disease symptoms, analyzing diagnoses, epidemiological research, book keeping, etc. The free text in these documents has been shown to be less ambiguous than text in general unrestricted documents (Ruch et al. 2001) making it feasible to successfully apply extraction techniques using tools from IE and NLP. Natural language processing has been used to analyze free text in

Proceedings of the NAACL HLT 2010 Second Louhi Workshop on Text and Data Mining of Health Documents, pages 8–14,
Los Angeles, California, June 2010.

medical domain for decision support (Chapman et al. 2005), classifying medical problem lists (Meystre and Haug 2005), extracting disease related information (Xu et al. 2004), building dynamic medications lists (Pakhomov et al. 2002), building applications for better data management, and for diagnosis detection. (Friedman et al. 2004, Roberts et al. 2008, Liu and Friedman 2004).

Our goal is to automatically generate diabetes report cards from the free text in physicians' letters. The report card can be used to detect populations at risk for diabetes mellitus and track their vital information over a period of time. Previous work in similar area has seen Turchin et al. (2005) identify patients with diabetes from the text of physician notes by looking for mention of diabetes and predefined list of medication names. They use a manually created list of negation tokens to detect false examples. They compare the process to manual chart review and billing notes and show the automatic system performs at par with manual review with the advantage of it being highly efficient.

In Turchin et al. (2006) the authors use regular expressions to extract blood pressure values and change of treatment for hypertension. They use a set of regular expressions to detect the presence of a blood pressure related tag, which predicts that the sentence is likely to contain a blood pressure value. The value itself is then extracted using regular expressions. They identify the strength of the process in it being relatively simple, efficient and quick to setup, while its weakness is its lack of generalization. Voorham and Denig (2007) solve a similar problem as in here and extract information regarding diabetes from free text notes using a number centric approach. They identify all positive numerical values and then attach respective labels to the values. They use a keyword based approach with a four word token window and apply a character sequence algorithm to check for spelling errors.

Extracting relevant information from free text represents a challenging problem since the task can be considered to be a form of reverse engineering and is above the mere presence of keywords or patterns. It is necessary to generate semantic representations to understand the text. The free text document may contain multiple values for the same label, and it's important to be able to distinguish and choose the correct value. These values could be:

- multiple readings (in which case a predefined rule may be enough, e.g. choosing the smallest mean arterial blood pressure value)
- potential target values (which may or may not be important)
- values taken over a period of time
- values taken at different locations
- values reflecting family history
- change in a value and not the actual value
- values influenced by some external reasons (e.g. take medication if the weight is above a certain value).

Friedman and Hripcsak (1999) discuss some of the many problems of dealing with free text in medical domain. One method to resolve these problems is to build a full grammar tree and assign semantic roles to accurately interpret the text. However, generating full parse trees for medical text requires specialized parsers developed for the clinical domain (Freidman, 2005). It has been shown that shallow syntactic approaches can yield similar results to the ones using full syntactic details (Gildea & Palmer, 2002).

In this work we use shallow syntactic and semantic features (manually created concept list and WordNet, Miller 1995) to tag information relating to the numerical values extracted from the text. We use machine learning tool WEKA (Hall et al. 2009) to build binary classifiers that pick positive values from the list of values extracted from the document. Our method allows us to build a robust and extendible system which should be easily portable to texts from different institutions and other medical domains.

2 Method

Our method extends Voorham's work in using the numeric value centered approach while developing a robust way to disambiguate between multiple values in the same document. The information extracted for the report card is divided into four categories: demographic information, numerical measurement values, medication list, and diagnoses. We currently have access to only one source of information, the free text in physicians' notes, hence all of the information needed for the report card is extracted from these notes. The extraction of demographic information is achieved using regular expressions/pattern matching based techniques. The demographic information extracted is year of birth, date of encounter and gender. The gender information is determined using a heuristic, which counts the number of third person masculine and feminine pronouns present in the text. Numerical measurement values extracted include blood pressure (systolic and diastolic), LDL, HDL, HbA1C, weight, total cholesterol, fasting glucose, glucose (unspecified) and creatinine. The medication list extraction process uses a manually created database of applicable medications. The diagnosis detection involves negation detection in the sentences that mention diabetes using the NegEx algorithm (Chapman et al. 2001).

In this study we use shallow syntactic and semantic attributes to build a system that extracts the physical examination and laboratory results data. The values are extracted as numeric value-label pairs. The system is divided into three main parts (Figure 1): preprocessing stage, extraction of the numeric value-label pairs, and testing the validity of the extracted pairs.

Preprocessing: The documents were originally stored in Microsoft Word format (WordML). They are converted to XML using XSLT transformation. All formatting information is stripped except for bold and italic font information and paragraph boundaries

The paragraphs in the document are further broken down into sentences and tokens. We use OPENNLP Maxent[1] library to do sentence boundary detection and tokenization. OPENNLP Maxent is based on maximum entropy algorithms described in Ratnaparkhi (1998) and Berger et al. (1996). The OPENNLP statistical tagger is used to assign syntactic tags to the tokens.

Data Extraction: In this phase the system extracts all potential numerical values and assigns them labels. The system loops through all of the tokens in the document, testing for numerical values. It tests each numerical token against a set of regular expressions and assigns them a list of potential labels based on the regular expression it matches. The system takes into account the presence of a measurement unit and revises the potential list of labels based on the unit. For each potential label, using a knowledge base, the system looks for concepts that validate the labels. The closest possible validation is accepted as pairing. The Edit distance algorithm is used to test for matching concepts in order to account for any spelling errors. The concepts are searched within the constraints of the sentence.

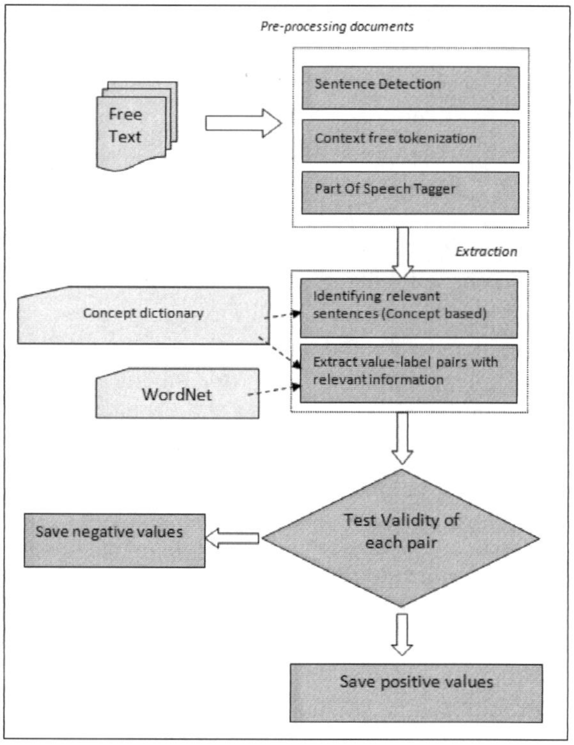

Figure 1 Process Flow diagram for the extraction process

In case multiple labels are validated because of the presence of multiple concepts in the same sentence, the label indicated by the closest concept is selected. For each pair, the system extracts a list of features which help to resolve for positive values. One exception to the sentence level boundary rule is: if no concepts are found in the sentence, and the sentence contains a third person singular inanimate pronoun, the search is extended to the previous sentence.

Testing Validity: The previous step extracts all possible label-numeric value pairs. As discussed earlier not all values are valid or of interest. In order to select positive values, binary classifiers were built for each label. The dataset used for training consisted of 900 documents (210 patients). The J48 (decision trees 4.5) and NBTree (Naïve Bayes decision trees) algorithms in WEKA were used to generate the machine learning classifiers.

[1] http://opennlp.sourceforge.net/

Features: The following is the list of features extracted for each pair.

a) Absolute distance between the label and the numerical value.
b) Label shared (Yes/No): Yes, if the same concept label is attached to another numerical value in the same document.
c) Closest verb token appearing left of the numerical value.
d) Presence of a modal verb (Yes/No)
e) Distance of numerical value from the modal verb (a positive value is assigned for the modal verb if it occurs before the numeric token, and a negative value when it appears after).
f) Conjunction present (Yes/No): If there is conjunction present between the label and numerical value or not.
g) Coreference present (Yes/No): If third person singular inanimate pronoun is present or not.
h) Negation concept present (Yes/No): True if there is any negation concept present in the vicinity of the numerical value/label. The negation concepts include not just negative statement markers, but also false cognates and other concepts collected by the domain experts. (e.g. systolic murmur or systolic volume do not indicate systolic pressure).
i) Locational Information token: The stemmed token is stored if it is recognized as a locational information token. The location information is deduced by generalizing each token and checking to see whether it resolves to one of many Locational cues in WordNet. The list of location indicators is presented in Figure 2. The cues are resolved against the WordNet hypernym definitions for that token.
j) Distance of numerical value from Locational token.
k) Temporal information token: Similar to (i), the stemmed token indicating temporal information. The temporal information token includes any tokens that indicate date or time. The list of temporal indicator cues in WordNet is shown in Figure 2.
l) Distance of the numerical value from the temporal token.

For features (c), (i) and (k) the tokens are stored in their uninflected form, achieved using Porter Stemmer. For the report card, in case of multiple positive values for the same label, the smallest value is selected. In the case of blood pressure, the smallest mean arterial pressure is selected.

Location
=>"*Facility, installation*" =>"*Housing, lodging, living accommodations*" =>"*Facility, installation*" =>"*Passage*" =>"*Structure, construction*" =>"*Road, route*" =>"*Geographic point, geographical point*" =>"*Location*"

Time
=>"*time period, period of time, period*" => "*time unit, unit of time*" => "*happening, occurrence, occurrent, natural event*"

Figure 2 WordNet hypernym based generalization cues for location and time indicators

3 Evaluation

Evaluation was done using a test set consisting of 804 documents from 260 patients (50 percent had positive diagnosis for diabetes). The test set was created by a first year student at Michael G Degroote School of Medicine at McMaster University. The reviewer manually analyzed the notes and extracted final values that would appear on the report card along with a time stamp for each value to indicate the source document. The human reviewer took approximately 10 minutes per patient; in comparison the computer analyzed the data at 6.43 patients per minute.

Evaluation results testing the performance of the system using the manually coded test set are shown in Table 1 below.

	Value	Precision	Recall	F-measure
1	Blood Pressure	98.2	96.9	97.8
2	LDL	96.4	94.2	95.3
3	HDL	100	98.3	99.1
4	Creatinine	97.2	92.1	94.5
5	Weight	95.6	92.9	94.2
6	TC	93.1	98.1	95.5
7	Glucose	90.7	85.7	87.7
8	F Glucose	88.8	80.0	84.2
9	HbA1C	90.9	86.9	88.8

Table 1 Precision/Recall for numerical values

4 Results and Discussion

The precision, recall and f-measure for all nine label values extracted for the system along with the recall values for the human reviewer are listed in Table 1. The system demonstrates high precision in extracting and selecting positive numeric value-label pairs. Blood pressure is extracted with a precision of 98.2% and recall 96.9%. HDL and LDL values are easy to spot and extract as they usually occur without description. At the lower end of precision are fasting glucose, glucose and HbA1C where precision results are in the range of 88-90%. The majority of errors for all categories occurred due to problems in identifying numeric values because of typing errors.

Figure 3 shows an example of the level of complexity resolved using the algorithm developed here. The clinical documents frequently have multiple values for weight and blood pressure in a single document. The lab values do not have the same level of multiplicity but it can occur. In this example, the extraction step extracts all five values, and the classifier successfully rejects values #3 and #5. To comply with the report card's output requirements the lowest mean arterial pressure of the remaining three values is adopted, which is the correct response. This approach is extendible to build a slot-filler system for the values, which would allow the system to reason on its choice.

In previous work, the disambiguation of the values is only based on the presence of negation concepts within a pre-specified boundary. We extend this to include a simple need based co-reference, location and temporal information, and a heuristic approach to include the head verb (it only takes into account the closest verb, which may or may not be the governing verb). The system can successfully detect negative values such as target values, previous values, change in value or values measured elsewhere.

The information extracted is stored in a structured MySQL database. The system allows multiple views on this information. Figure 4 shows the output for blood pressure and creatinine for a patient that was created from the information extracted from the free text.

Blood pressure initially was 196/92 in the left arm and 194/90 in the right arm. Usually, the patient states, at home it is 140-150/80. The blood pressure subsequently decreased to 160/88 when I waited several minutes and had him calm down....... Target blood pressure should be below 140/90.

Values Extracted:

1) 196/92 = {Loc: left arm, Verb: was}
2) 194/90 = {Loc: right arm, Verb: was, Conjunction: True}
3) 140-150/80 = {Loc: home, Co-ref: It, Verb: is}
4) 160/88 = {Verb: Decreased}
5) 140/90 = {Verb: be, Modal: should}

Figure 3 Example 1

At this time we have not evaluated the contribution of each feature individually, as this requires building a comprehensive test set; it remains as future work.

5 Conclusion

Our preliminary results demonstrate that the system performs with high precision and recall at the task of extracting numerical values. It also shows the ability to build a patient-chart abstractor within the restricted domain. The use of semantic and syntactic features enables the system to tag the values which permit the overall extraction process to generate more informative numeric value-label pairs. The use of machine learning algorithms coupled with a large enough learning dataset produces a robust system that should work reliably on similar data from any source. We plan to test the system on a dataset obtained from the free text notes of endocrinologists at a different health institution to validate the generalization of the algorithm. The next step for the Diabetes Report Card is to extract the list of medications and track any changes in medication, dosage and frequency.

Acknowledgments

A special thanks to Michael Domenic Corbo for doing the manual review and creating the gold standard dataset.

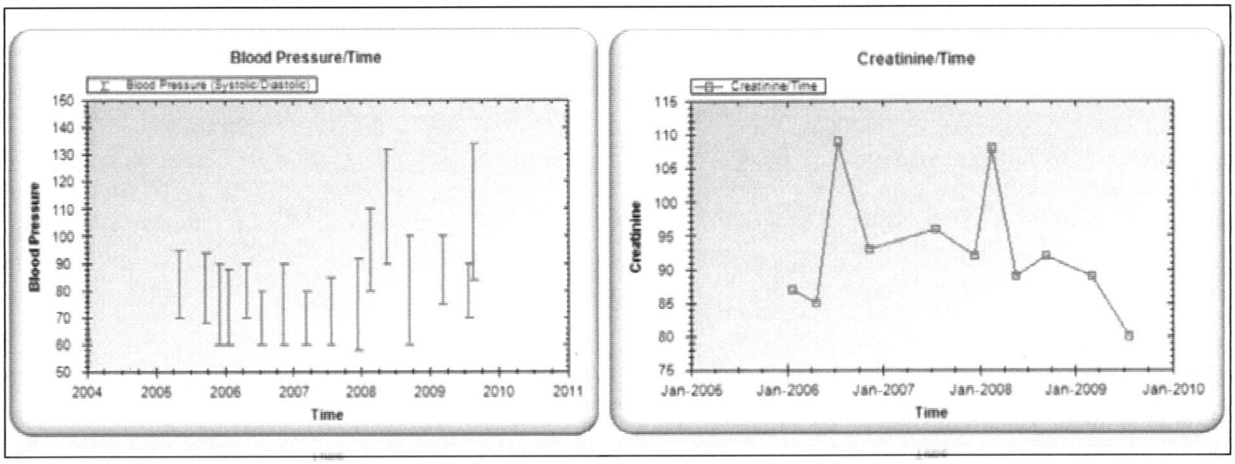

Figure 4 System Output: Automatically generated graphs for blood pressure and creatinine values for a patient

6 References

Berger, A. L., Pietra, V. J., & Pietra, S. A. (1996). A maximum entropy approach to natural language processing. *Computational Linguistics* , 39-71.

Chapman, W. W., Christensenb, L. M., Wagnera, M. M., Haugb, P. J., Ivanova, O., Dowlinga, J. N., et al. (2005). Classifying free-text triage chief complaints into syndromic categories with natural languages processing. *Artificial Intelligence in Medicine* , 31-40.

Chapman, W., Bridewell, W., Hanbury, P., Cooper, G., & Buchanan, B. (2001). Evaluation of negation phrases in narrative clinical reports. *Proc AMIA Symp* , 105-114.

Freidman, C. (2005). Semantic Text Parsing for Patient Records. In *Medical Informatics* (pp. 423-448). Springer US.

Friedman, C., & Hripcsak, G. (1999). Natural Language Processing and Its Future in Medicine. *Acad Med* , 890-895.

Friedman, C., Shagina, L., Lussier, Y., & Hripcsak, G. (2004). Automated Encoding of Clinical Documents based on Natural Language Processing. *Journal of American Medical Informatics Association* .

Gildea, D., & Palmer, M. (2002). The Necessity of Syntactic Parsing for Predicate Argument Recognition. *Association for Computational Linguistics*, (pp. 239-246).

Hall, M., Frank, E., Holmes, G., Pfahringer, B., Reutemann, P., & Witten, I. H. (2009). The WEKA Data Mining Software: An Update. *SIGKDD Explorations* .

Heinze, D. T., Morsch, M. L., & Holbrook, J. (2001). Mining free-text medical records. *AMIA*, (pp. 254-258).

Hripcsak, G., Friedman, C., Alderson, P., DuMouchel, W., Johnson, S., & Clayton, P. (1995). Unlocking clinical data from narrative reports: a study of natural language processing. *Ann Intern Med* , 681-689.

Liu, H., & Friedman, C. (2004). CliniViewer: a tool for viewing electronic medical records based on natural language processing and XML. *MedInfo* , 639-643.

Lovis, C., Baud, R. H., & Plancheb, P. (2000). Power of expression in the electronic patient record: structured data or narrative text? *International Journal of Medical Informatics* , 101-110.

Mcdonald, C. J. (1997). The Barriers to Electronic Medical Record Systems and How to Overcome Them. *Journal of the American Medical Informatics Association* , 213-221.

Meystre, S., & Haug, P. J. (2005). Automation of a problem list using natural language processing. *BMC Medical Informatics and Decision Making* , 5-30.

Miller, G. A. (1995). WordNet: A Lexical Database for English. *Communications of the ACM , 38*, 39-41.

Pakhomov, S. V., Ruggieri, A., & Chute, C. G. (2002). Maximum entropy modeling for mining patient medication status from free text. *Proceedings of the American Medical Informatics*, (pp. 587–591).

Ratnaparkhi, A. (1998). *Maximum Entropy Models for Natural Language Ambiguity Resolution.* Phd Thesis.

Roberts, A., Gaizauskas, R., Hepple, M., & Guo, Y. (2008). Mining clinical relationships from patient narratives. *Natural Language Processing in Biomedicine (BioNLP) ACL Workshop.*

Ruch, P., Baud, R., Geissbuhler, A., & Rassinoux, A.-M. (2001). Comparing general and medical texts for information retreival based on natural language processing: An inquiry into lexical disambiguation., (pp. 261-266).

Turchin, A., Kohane, I., & Pendergrass, M. (2005). Identification of patients with diabetes from the text of physician notes in the electronic medical record. *Diabetes Care* , 1794-1795.

Turchin, A., Kolatkar, N., Grant, R. W., Makhni, E. C., Pendergrass, M. L., & Einbinder, J. S. (2006). Using regular expressions to abstract blood pressure and treatment intensification information from the text of physician notes. *Journal of the American Medical Informatics Association* , 691-696.

Voorham, J., & Denig, P. (2007). Computerized Extraction of Information on the Quality of Diabetes Care from Free Text in Electronic Patient Records of General Practitioners. *The Journal of the American Medical Informatics Association* , 349-354.

Xu, H., Anderson, K., Grann, V. R., & Friedman, C. (2004). Facilitating Research in Pathology using Natural Language Processing. *Proc AMIA Symp*, (p. 1057).

Negation Detection in Swedish Clinical Text

Maria Skeppstedt
DSV/Stockholm University
Forum 100
SE-164 40 Kista, Sweden
mariask@dsv.su.se

Abstract

NegEx, a rule-based algorithm that detects negations in English clinical text, was translated into Swedish and evaluated on clinical text written in Swedish. The NegEx algorithm detects negations through the use of trigger phrases, which indicate that a preceding or following concept is negated. A list of English trigger phrases was translated into Swedish, taking grammatical differences between the two languages into account. This translation was evaluated on a set of 436 manually classified sentences from Swedish health records. The results showed a precision of 70% and a recall of 81% for sentences containing the trigger phrases and a negative predictive value of 96% for sentences not containing any trigger phrases. The precision was significantly lower for the Swedish adaptation than published results on the English version, but since many negated propositions were identified through a limited set of trigger phrases, it could nevertheless be concluded that the same trigger phrase approach is possible in a Swedish context, even though it needs to be further developed.

1 Introduction

Medical documentation, such as patient records, is today often stored in a digital, searchable format. This opens the possibility of extracting information, which for example could be used for disease surveillance or to find new, unknown connections between patient background, symptoms and diseases. When extracting information from a text, it is not only the words that occur in the text that are important, but also whether these words are negated or not. This is especially true when it comes to patient records, since when describing the status of a patient, the physician often reasons by excluding various possible diagnoses and symptoms.

Most work on detecting negations in medical language has been carried out for English, and very little has been carried out for other languages, as for example Swedish. This article will therefore focus on the task of finding whether a concept in a clinical text written in Swedish is negated or not.[1]

2 Related research

There are many different methods for detecting whether a concept is negated. Rokach et al. (2008) give a good overview of some approaches for detecting negations. The methods can be divided into two main groups; knowledge engineering methods and machine learning methods. Knowledge engineering methods have the advantage that a large annotated corpus is not needed, but the disadvantage that rules have to be written manually, which is often time-consuming. Negation detection based on machine learning methods, on the other hand, is faster to implement and often works better when a text is not completely grammatical, which is often the case with clinical texts. (Rokach et al., 2008)

Since little previous work has been done on negation detection in Swedish medical text, the first step

[1] This research has been carried out after approval from the Regional Ethical Review Board in Stockholm (Etikprövningsnämnden i Stockholm), permission number 2009/1742-31/5.

Proceedings of the NAACL HLT 2010 Second Louhi Workshop on Text and Data Mining of Health Documents, pages 15–21,
Los Angeles, California, June 2010.

for Swedish negation detection is to adapt a simple knowledge engineering method that is used for detecting negations in English, an algorithm called NegEx. (Chapman et al., 2001b)

2.1 The NegEx algorithm

NegEx detects *pertinent negatives* in English patient records, that is "findings and diseases explicitly or implicitly described as absent in a patient". Given a sentence and a chosen *proposition* in this sentence, NegEx determines if that proposition is negated or not. An example would be *"Extremities showed no cyanoses."*, in which the proposition is *cyanoses*. (Chapman et al., 2001b)

The NegEx algorithm uses regular expressions and three lists of phrases. The first list, the pre-negation list, consists of trigger phrases which indicate that a proposition that follows them is negated in the sentence, for example *no signs of*. The second list, the post-negation list, consists of trigger phrases that indicate that a proposition preceding them is negated, as the phrase *unlikely*. Finally, the third list consists of pseudo-negation phrases, phrases that are similar to negation triggers, but that do not trigger negation, for example *not certain if*. The algorithm judges the proposition to be negated if it is in the range of one to six words from a post- or pre-negation trigger. (Chapman et al., 2001b)

NegEx has later been further developed into *NegEx version 2*[2], for example through the addition of more triggers and by limiting the scope of the negation through a list of conjunctions.

In the evaluation of NegEx, the propositions consisted of UMLS[3] phrases that belonged to any of the UMLS categories *finding*, *disease or syndrome* or *mental or behavioral dysfunction* and that could also be found in the describing text of an ICD-10 code[4]. Sentences containing these UMLS phrases were extracted from discharge summaries. Thereafter, 500 of the extracted sentences that contained at least one negation trigger and 500 sentences that did not contain a negation trigger were randomly selected. A few sentences that contained phrases that were suspected to sometimes indicate a negation, but that were not in the three lists, were included in the first group. The sentences were then categorised by physicians into containing an *affirmed proposition*, a *negated proposition* or an *ambiguous proposition*. The inter-rater agreement was almost 100%. For the NegEx evaluation, the categories *affirmed* and *ambiguous* were grouped into the category *not negated*. The results showed a precision of 84% and a recall of 82% for sentences in the group with negation triggers and a negative predictive value of 97% for sentences in the group without triggers. Of the correctly found negations, 82% were triggered by only three negation triggers; *no*, *without* and *no evidence of*. Moreover, only 15 of the 35 negation triggers were found in the test set. The trigger *not* had a precision of 58%, which was much lower than the precision for the other common triggers. (Chapman et al., 2001b)

An evaluation of the NegEx algorithm on ten other kinds of reports has also been carried out. The average precision of NegEx was 97%, and 90% of the detected negations were triggered by only seven negation phrases, with the four most frequent being *no*, *denies*, *without* and *no evidence*. (Chapman et al., 2001a)

In a later study by Goldin and Chapman (2003), a Naive Bayes classifier and a decision tree were used to classify which occurrences of the trigger *not* that indicated a negation, based on features such as surrounding words and their part of speech. Both these methods resulted in an increased precision.

3 Research Question

An evaluation was carried out on how the NegEx algorithm performs on health records written in Swedish, compared to health records written in English. The hypothesis was that the results for Swedish would be similar to the results for English, since the two languages are grammatically close. This comparison could give an indication of whether it is possible to adapt more advanced methods for negation detection into Swedish, and the results could also be used as a baseline for comparing the results of other methods.

4 Translation and adaption method

In order to use NegEx on a Swedish text, there must be a list of Swedish phrases that trigger negation.

[2] http://www.dbmi.pitt.edu/chapman/negex.html

[3] See Bodenreider (2004) for a description of UMLS

[4] http://www.who.int/classifications/icd/en/

4.1 Translating trigger phrases

The triggers for Swedish were obtained by translating the phrases for *NegEx version 2*. The translations were made with the help of a web-based English-Swedish dictionary[5] and with the help of Google translate[6]. In the cases where there was a good translation neither in the dictionary nor in the Google translation, the negation was translated by the author of this article. When it was not possible to find a good Swedish translation, the phrase was omitted. A total of 148 phrases were translated. Almost all negation phrases were general English terms. However, in a few cases they consisted of specific medical terms, and in these cases the translation was made by a physician. In many instances the dictionary offered many translations, and in other cases the same translation was offered for different English phrases. In the cases where several translations were offered, all of them were added to the list of Swedish negations.

4.2 Expanding the translated trigger phrases

English and Swedish are both Germanic languages (Crystal, 1997) and they have a similar grammar. Nevertheless, there are some grammatical differences that have to be taken into account through an expansion of the list of translated trigger phrases.

Swedish has two grammatical genders (common gender and neuter gender), whereas the English language lacks grammatical gender. Adjectives and some quantifiers in Swedish have a gender concord, as well as a number concord (Dahl, 1982). To compensate for this, the English negative quantifier *no* was translated into three different forms of the corresponding Swedish negative quantifier, namely *inga*, *ingen* and *inget*. Inflections of all adjectives in the trigger phrases were also generated. This was accomplished by invoking the Granska inflector[7].

The English combinations of aspect and tense do not always correspond directly to a Swedish verb form (Dahl, 1982). Therefore, a direct translation of the different forms of a verb in the trigger phrase list was not performed. The lemma form of the verb was instead added to the list of negation triggers in Swedish, and from this all inflections of the verb were generated, again using the Granska inflector.

The difference connected with the do-construction did not need to be taken into account. When negating a non-auxiliary verb in English, the do-construction is used. This type of construction does not exist in Swedish. The phrase *han vet* (*he knows*) would for example be negated as *han vet inte* (*he knows not*) (Svartvik and Sager, 1996). However, the NegEx algorithm only checks if the proposition is less than six words to the right of the word *inte* (*not*), and when it is, it will consider the proposition to be negated. The lack of a do-construction should therefore not affect the results.[8]

Swedish has a word order inversion in subordinate clauses. The position of the negating adverb is changed, and it is instead positioned immediately before the verb (Holmes and Hinchliffe, 2008). When stressing the negation, there is also the possibility of using this word order in the main clause (Sells, 2000). A version with reversed word order was therefore generated for trigger phrases containing some of the most common adverbs. From the translation of the trigger phrase *has not*, a version with the word order *not has* was for example generated.

The frequency of the Swedish trigger phrases was counted on a text other than the test set, and the most frequent trigger phrases were selected. The number of selected phrases was two more than used in the English NegEx evaluation, to compensate for Swedish gender and number concord[9].

5 Evaluation method

5.1 Construction of test data

Propositions to use for evaluating the performance of the Swedish version of NegEx were taken from the Swedish translation of the ICD-10 codes. However, the description in the ICD-10 code list often contains both the name of a symptom or disease and a clarification or specification of it, which has the

[5]http://www.norstedtsord.se

[6]http://translate.google.com

[7]http://www.csc.kth.se/tcs/humanlang/tools.html

[8]When negating the actual verb on the other hand, the position of the word *not* is different in English and Swedish. In order for the Swedish NegEx to handle verb phrase propositions, this difference has to be accounted for.

[9]The triggers that were used can be downloaded from http://people.dsv.su.se/~mariask/resources/triggers.txt

effect that simple string matching would not find some of the most common symptoms and diseases. An automatic pre-processing of the ICD-10 code list was therefore first accomplished, where for example text within parenthesis and clarifications such as *not specified* or *other specified forms* were removed. To find more names of symptoms and diseases, additional lists were also added, including the KSH97-P[10], an adaption of the ICD-10 codes for primary care, and the MeSH terms under the sections *diseases* and *mental disorders*.

The test data was extracted from a set of sentences randomly chosen from the assessment part of Swedish health records from the Stockholm EPR Corpus (Dalianis et al., 2009). From this set, sentences that contained any of the propositions in the proposition list were extracted, also when the proposition was part of a compound word. Neither the preprocessing of the ICD-10 code list nor the detection of a proposition in a compound word was perfect and therefore some words that were not comparable with *findings*, *diseases or syndromes* or *mental or behavioral dysfunctions*, were added to the list of propositions. Sentences containing these were manually filtered out from the test data.

The chosen sentences were ordered in a list of pairs, consisting of the sentence and the proposition. If a sentence contained more than one proposition, the sentence was added to the list one time for each proposition.

In order to be able to compare the English and Swedish versions of NegEx, the same evaluation method was used, and two groups of test sentences were constucted. The first group, *Trig*, contained 202 sentences with at least one of the trigger phrases. The second group, *Non-Trig*, contained 234 sentences without any of the trigger phrases.

5.2 Classification of test data

The propositions were manually classified into the categories *affirmed*, *negated* and *ambiguous* by a rater without medical education. The categories *affirmed* and *ambiguous* were thereafter collapsed into the category *not negated*. The results are presented in Table 1.

Of the 202 sentences in group *Trig*, 70 were also classified by a physician. The inter-rater agreement between the physician and the other rater with respect to the two groups *negated* and *not negated* was 80%.

The majority of the sentences where there was disagreement were judged as negated by the physician rater and ambiguous by the other rater, or ambiguous by the physician rater and negated by the other rater. There was no evident systematic tendency to judge the propositions as more or less ambiguous by either of the two raters.

When there was a difference in opinion of how to classify the proposition, the classification made by the physician was chosen. Also sentences that were subjectively judged by the rater as not possible to rate without deep medical knowledge, were rated by the physician.

	Negated	Not negated	Total
Trig	90	112	202
Non-Trig	10	224	234

Table 1: Number of sentences manually classified as *negated* and *not negated* for each of the groups *Trig* and *Non-Trig*. Group *Trig* only contains sentences with trigger phrases and Group *Non-Trig* only contains sentences without trigger phrases.

6 Results

The Swedish version of NegEx was executed with the sentences in group *Trig* and the sentences in group *Non-Trig* as input sentences.[11] As shown in Table 2, group *Trig* had a precision of 70% and a recall of 81%. Group *Non-Trig* had a negative predictive value of 96%, as shown in Table 3.

When comparing Swedish and English results for recall using the χ^2-test, no significant difference was found between them. (p-value $>> 0.1$). When comparing the results for precision using the χ^2-test, it was significantly lower for Swedish. ($p < 0.001$).

The precision of each trigger was also counted and the results are shown in Table 4.

[10]http://www.socialstyrelsen.se/publikationer1996/1996-4-1

[11]http://code.google.com/p/negex/updates/list is the web location of NegEx (negex.python.zip, 2009). NegEx could be used in a Swedish context without any major modifications.

Group *Trig*	English	Swedish
recall (sensitivity)	82.00 %	81 %
specificity	82.50 %	71 %
precision (ppv)	84.49 %	70 %
npv	80.21 %	82 %

Table 2: Group *Trig*, 500 English sentences and 202 Swedish sentences. *Recall*: No. of correctly detected negated propositions divided by no. of manually rated negated propositions. *Specificity*: No. of propositions correctly detected as not negated divided by no. propositions that were manually rated as not negated. *Precision*: No. of correctly detected negated propositions divided by total no. of propositions that NegEx classified as negated. *Negative predictive value*: No. of propositions that NegEx correctly did not classify as negated divided by total no. of propositions that NegEx did not classify as negated. (Figures for English from Chapman et al. (2001b).)

Group *Non-Trig*	English	Swedish
npv	96.99 %	96 %

Table 3: Group *Non-Trig*, 500 English sentences and 234 Swedish sentences. (Figures for English from Chapman et al. (2001b).)

7 Discussion

The comparison between the English and Swedish evaluations is complicated by the fact that the Swedish test data had lower inter-rater agreement, which adds uncertainty to the Swedish results. This difference could perhaps be partly explained by the different types of health records; the English version was evaluated on discharge summaries, whereas the Swedish version was evaluated on the assessment part of a health record, which possibly contains more reasoning and thereby perhaps more ambiguous expressions.

Also, the fact that group *Trig* in the evaluation of the English version also included some sentences not containing trigger phrases complicates the comparison.

It could, however, be concluded that the precision is lower for Swedish. The following error types could at least account for some of this difference:

It is difficult to draw a line between what is an ambiguous expression and what is a negation, both for the raters and for the NegEx program. The above-mentioned difference in type of evaluation data could have resulted in lower precision and recall for the Swedish version.

Phrase		Precision	Occur.
inga tecken	(no signs of)	89 %	9
ingen	(no)	89 %	27
ej	(not)	75 %	8
inga	(no, plural)	67 %	15
utan	(without)	63 %	8
inte har	(not have)	60 %	5
inte	(not)	57 %	21
icke	(non-, not)	0 %	4

Table 4: The most frequent triggers, their precision and the number of times they occur in the sentences.

It is a common construction for a name of a disease, or a version of a disease, to have a name that starts with the word *icke* (*non-, not*), for example *icke allergisk astma*. The disease is present in the patient, even though the word *icke* is interpreted as a negation trigger by NegEx. In the test data, all the occurrences of the word *icke* are constructions like this, thus having a negative impact on precision.

The Swedish word for *without* (*utan*) has a double meaning. It is also a conjunction meaning *but*. This gives rise to a few instances where the program incorrectly classifies a proposition as negated, resulting in lower precision.

Other error types were also identified. These were, however, not specific for Swedish or for the type of test data, and could therefore not account for the difference in precision between the English and Swedish versions of NegEx. Examples are when the negation of the proposition occurs in a conditional clause or when the scope of the trigger should be less than the NegEx scope of six words, for example when the scope is limited by a conjunction.

7.1 Identified negation triggers

In the test set, only 16 of the 39 negation triggers were found, and among them, only 12 correctly negated a proposition. This is close to the English version where 15 of 37 triggers were found. None of the post-negation triggers were found in the Swedish test data.

In the English version of NegEx, 82% of the correctly found negations were triggered by the three negation phrases *no*, *without* and *no evidence of*. In the Swedish version, the three most common triggers were the common gender version of *no* (*ingen*), *not* (*inte*) and the plural form of *no* (*inga*). Together, they constitute 63% of the total number of correctly identified negations. If the trigger in fourth place, *no signs of*, is also counted, they make up 75% of the correctly negated propositions. In both English and Swedish there are thus a small number of negation triggers that are very common.

It can also be noted that both in Swedish and English, the precision of the trigger *not* (*inte*) is low.

No other common negation triggers were found in the test data. The only re-occurring trigger that was not included in any of the three lists were different forms of the phrase *rule out*.

8 Conclusion

The Swedish version of the NegEx algorithm had a significantly lower precision than the English version, and for the recall no significant conclusions could be drawn. Not taking the uncertainty of the low inter-rater agreement into account, the Swedish version has a precision of 70% and a recall of 81% for sentences containing the trigger phrases and a negative predictive value of 96% for sentences not containing any trigger phrases. As for the English version, a small number of trigger phrases accounted for the majority of detected negations.

Since a limited set of triggers can be used to identify many negations also in Swedish, this simple approach of the NegEx algorithm can be used as a base method for identifying negations in Swedish. However, even for use in a system without high demands on robustness, the method needs to be further developed.

From the relatively low inter-rater agreement, especially with respect to concepts that might be classified as either ambiguous or negated, it can be concluded that it is a difficult task also for a human rater to determine what is an ambiguity expressed as a negation or an actual negation.

9 Limitations

The most important limitation of this study is the relatively low inter-rater agreement, and the fact that most of the sentences were rated by a person who did not have a medical education. The lack of medical knowledge may have lead to mistakes when classifying the test data and could probably also partly explain the low inter-rater agreement.

Another limitation is that errors in the module for selecting sentences lead to that a few test sentences did not contain a symptom, disease or equivalent. Consequently, these sentences had to be filtered out manually.

As in the study by Chapman et al. (2001a), no analysis has been made of the occurrences of negations that stretch over sentence boundaries.

10 Future work

To automatically distinguish an ambiguity from a negation is not always trivial. However, the errors originating from the other error types mentioned could be limited through the use of more advanced natural language processing methods. The cases where the phrase *icke* does not trigger a negation, could probably be detected by a simple regular expression rule. Which meaning of the phrase *utan* that is intended could perhaps be detected by the machine learning methods used by Goldin and Chapman (2003). A list of conjunctions that limit the scope of the negations, as in *NegEx version 2*, could also be used to increase the precision, and a similar method could be used to detect when the proposition is negated in a conditional phrase.

It would also be interesting to use the complete list of negation triggers that was constructed for this study, instead of limiting the size to that of the NegEx trigger list, and to evaluate this list on a larger test set. This evaluation could also determine whether there are any common Swedish negation triggers that were not obtained by translating the English trigger list.

Acknowledgments

I would like to thank my supervisors Hercules Dalianis and Gunnar Nilsson for valuable comments on this paper, and specifically Gunnar for the help with the classification of the sentences. I would also like to thank Birgitta Melin Skeppstedt for initial help with the statistical calculations and Sumithra Velupillai for the support on the early stages of the work. Many thanks also to the three anonymous reviewers of the paper.

References

Olivier Bodenreider. 2004. The unified medical language system (umls): integrating biomedical terminology. *Nucleic Acids Res*, 1;32(Database issue).

Wendy W Chapman, Will Bridewell, Paul Hanbury, Gregory F. Cooper, and Bruce G. Buchanan. 2001a. Evaluation of negation phrases in narrative clinical reports. *Proc AMIA Symp*, pages 105–109.

Wendy W. Chapman, Will Bridewell, Paul Hanbury, Gregory F. Cooper, and Bruce G. Buchanan. 2001b. A simple algorithm for identifying negated findings and diseases in discharge summaries. *J Biomed Inform*, 34(5):301–310, Oct.

David Crystal. 1997. *The Cambridge encyclopedia of language*. Cambridge University Press, second edition.

Östen Dahl. 1982. *Grammatik*. Studentlitteratur.

Hercules Dalianis, Martin Hassel, and Sumithra Velupillai. 2009. The Stockholm EPR Corpus - Characteristics and Some Initial Findings. In *Proceedings of ISHIMR 2009, Evaluation and implementation of e-health and health information initiatives: international perspectives. 14th International Symposium for Health Information Management Research, Kalmar, Sweden*, pages 243–249.

Ilya M. Goldin and Wendy W. Chapman. 2003. Learning to detect negation with 'not' in medical texts. *ACM SIGIR '03 Workshop on Text Analysis and Search for Bioinformatics: Participant Notebook, Acknowledgements Toronto, Canada: Association for Computing Machinery;.*

Philip Holmes and Ian Hinchliffe. 2008. *Swedish: An Essential grammar*. Routledge.

Lior Rokach, Roni Romano, and Oded Maimon. 2008. Negation recognition in medical narrative reports. *Information Retrieval*, 11(6):499–538, December.

Peter Sells. 2000. Negation in Swedish: Where it's not at. In *Online Proceedings of the LFG-00 Conference. Stanford: CSLI Publications. (At http://csli-publications.stanford.edu/LFG/5/lfg00.html).*

Jan Svartvik and Olof Sager. 1996. *Engelsk universitetsgrammatik*. Liber.

Using Domain Knowledge about Medications to Correct Recognition Errors in Medical Report Creation

Stephanie Schreitter
Alexandra Klein
Johannes Matiasek
Austrian Research Institute
for Artificial Intelligence (OFAI)
Freyung 6/6
1010 Vienna, Austria
firstname.lastname@ofai.at

Harald Trost
Section for Artificial Intelligence
Center for Med. Statistics, Informatics,
and Intelligent Systems
Medical University of Vienna
Freyung 6/2
1010 Vienna, Austria
harald.trost@meduniwien.ac.at

Abstract

We present an approach to analysing automatic speech recognition (ASR) hypotheses for dictated medical reports based on background knowledge. Our application area is prescriptions of medications, which are a frequent source of misrecognitions: In a sample report corpus, we found that about 40% of the active substances or trade names and dosages were recognized incorrectly. In about 25% of these errors, the correct string of words was contained in the word graph. We have built a knowledge base of medications based on information contained in the Unified Medical Language System (UMLS), consisting of trade names, active substances, strengths and dosages. From this, we generate a variety of linguistic realizations for prescriptions. Whenever an inconsistency in a prescription is encountered on the best path of the word graph, the system searches for alternative paths which contain valid linguistic realizations of prescriptions consistent with the knowledge base. If such a path exists, a new concept edge with a better score is added to the word graph, resulting in a higher plausibility for this reading. The concept edge can be used for rescoring the word graph to obtain a new best path. A preliminary evaluation led to encouraging results: in nearly half of the cases where the word graph contained the correct variant, the correction was successful.

1 Introduction

Automatic speech recognition (ASR) is widely used in the domain of medical reporting. Users appreciate the fact that the records can be accessed immediately after their creation and that speech recognition provides a hands-free input mode, which is important as physicians often simultaneously handle documents such as notes and X-rays (Alapetite et al., 2009). A drawback of using ASR is the fact that speech-recognition errors have to be corrected manually by medical experts before the resulting texts can be used for electronic patient records, quality control and billing purposes. This manual post-processing is time-consuming, which slows down hospital workflows.

A number of recognition errors could be avoided by incorporating explicit domain knowledge. We consider prescriptions of medications a good starting point as they are common and frequent in the various medical fields. Furthermore, they contain trade names and dosages, i.e. proper names and digits, which are frequently misrecognized by ASR in all domains.

For our approach, we have extracted and adapted information about medications from the Unified Medical Language System (UMLS) (Lindberg et al., 1993). This data contains information about trade names, active substances, strengths and dosages and can easily be modified, e.g. when new medications are released.

In the first step, we assessed the potential for improvement by analyzing a sample corpus of medical reports. It turned out that in 4383 dictated reports which were processed by a speech-recognition system, the word-error rate for medications was about 40%, which is slightly higher than the the average word-error rate of the reports. Examining a sample

Proceedings of the NAACL HLT 2010 Second Louhi Workshop on Text and Data Mining of Health Documents, pages 22–28,
Los Angeles, California, June 2010.

of word graphs for the reports, we realized that in about 30% of these errors, the correct string of words was contained in the word graph, but not ranked as the best path.

In the following sections, we will first give an overview of previous approaches to detecting speech-recognition errors and semantic rescoring of word-graph hypotheses. Then, we will describe how we have adapted information about medications from the UMLS to enhance the word graph with concept nodes representing domain-specific information. Finally, we will illustrate the potential for improving the speech-recognition result by means of an evaluation of word graphs for medical reports which were processed by our system.

2 Extraction of Medication Information, Error Handling and Semantic Rescoring

(Gold et al., 2008) gives an overview on extracting structured medication information from clinical narratives. Extracted medication information may serve as a base for quality control, pharmaceutical research and the automatic creation of Electronic Health Records (EHR) from clinical narratives. The *i2b2* Shared Task 2009 focussed on medication extraction, e.g. (Patrick and Li, 2009; Halgrim et al., 2010). These approaches work on written narrative texts from clinical settings, which may have been typed by physicians, transcribed by medical transcriptionists or recognized by ASR and corrected by medical transcriptionists.

In contrast, our approach takes as input word graphs produced by an ASR system from dictated texts and aims at minimizing the post-processing required by human experts.

Speech-recognition systems turn acoustic input into word graphs, which are directed acyclic graphs representing the recognized spoken forms and their confidence scores (Oerder and Ney, 1993). In most speech-recognition systems, meaning is implicitly represented in the language model (LM), indicating the plausibility of sequences of words in terms of n-grams. It has often been stated that the introduction of an explicit representation of the utterance meaning will improve recognition results. Naturally, this works best in limited domains: the larger an application domain, the more difficult it is to build an optimal knowledge representation for all possible user utterances. Limited domains seem to be more rewarding with regard to coverage and performance. Consequently, combining speech recognition and speech understanding has so far mostly resulted in applications in the field of dialogue systems where knowledge about the domain is represented in terms of the underlying database, e.g. (Seneff and Polifroni, 2000).

Several approaches have investigated the potential of improving the mapping between the user utterance and the underlying database by constructing a representation of the utterance meaning. Meaning analysis is either a separate post-processing step or an integral part of the recognition process. In some approaches, the recognition result is analyzed with regards to content to support the dialogue manager in dealing with inconsistencies (Macherey et al., 2003). As far as dictated input is concerned, which is not controlled by a dialogue manager, (Voll, 2006) developed a post-ASR error-detection mechanism for radiology reports. The hybrid approach uses statistical as well as rule-based methods. The knowledge source UMLS is employed for measuring the semantic distance between concepts and for assessing the coherence of the recognition result.

In other approaches, the analysis of meaning is integrated into the recognition process. Semantic confidence measurement annotates recognition hypotheses with additional information about their assumed plausibility based on semantic scores (Zhang and Rudnicky, 2001; Sarikaya et al., 2003). (Gurevych and Porzel, 2003; Gurevych et al., 2003) present a rescoring approach where the hypotheses in the word graph are reordered according to semantic information. Usually, conceptual parsers are employed which construct a parse tree of concepts representing the input text for mapping between the recognition result and the underlying representation. Semantic language modeling (Wanget al., 2004; Buehler et al., 2005) enhances the language model to incorporate sequences of concepts which are considered coherent and typical for a specific context. In these approaches, the representations of the underlying knowledge are created specially for the applications or are derived from a text corpus.

In our approach, we aim at developing a prototype

for integrating available knowledge sources into the analysis of the word graph during the recognition process. We have decided not to integrate the component directly into the ASR system but to introduce a separate post-processing step for the recognition of information about medications with the word graphs as interface. This makes it easier to update the medication knowledge base, e.g. if new medications are released. Furthermore, it is not necessary to retrain the ASR system language model for each new version of the medication knowledge base.

3 Knowledge Base and Text Corpus

For our approach, we prepared a knowledge base concerning medications and dosages, and we used a corpus of medical reports, dictated by physicians in hospitals. The ASR result and a manual transcription is available for each report. For a subset of the corpus, word graphs could be obtained. By aligning the recognition result with the manual transcriptions, error regions can be extracted.

3.1 Knowledge Base

As it is our aim to find correct dosages of medications in the word graph, we built a domain-specific knowledge base which contains medications and strengths as they occur in prescriptions. In our sample of medical reports, about 1/3 of the medications occurred as active ingredients while the rest were trade names. Therefore, both had to be covered in our knowledge base which is based on RxNorm (Liu et al., 2005). RxNorm is a standardized nomenclature for clinical drugs and drug delivery devices and part of UMLS, ensuring a broad coverage of trade names and active ingredients. Of several available versions of RxNorm, the semantic branded drug form is the most suitable one for our purposes as it contains pharmaceutical ingredients, strengths, and trade names. For example, the trade name *Synthroid®* is listed as follows:

Thyroxine 0.025 MG Oral Tablet [Synthroid®]

Thyroxine is the active ingredient with the dosage value 0.025 and the dosage unit milligrams. The dosage unit form is oral tablet.

We used a RxNorm version with 1,508 active substances and 7,688 trade names (11,263 trade names counting the different dosages). The active ingredients in RxNorm are associated with Anatomical Therapeutic Chemical (ATC) Codes.

3.2 Sample Corpus

The corpus is a random sample of 924 clinical reports which were dictated by physicians from various specialties and hospitals. The dications were processed by an ASR system and transcribed by human experts. Word graphs marked with the best path (indicating the highest acoustic and language-model scores) represent the recognition result. Tradenames are part of the recognition lexicon, but they are frequently misrecognized.

Of the 9196 medications (i.e. trade names and active substances) in RxNorm, only 330 (3.6%) appeared in the sample corpus.

We searched the corpus for recognition errors concerning trade names, active ingredients and their dosages by comparing the manual transcriptions to the best paths in the word graphs, and a list of the mismatches (i.e. recognition errors) and their frequencies was compiled. It turned out that 39.3% of all trade names and active ingredients were recognized incorrectly. The average ASR word-error rate of the reports was 38.1%. Aproximately 1-2% of the trade names were not covered by RxNorm.

4 Approach

Our approach consists of a generation mechanism which anticipates possible spoken forms for the content of the knowledge base. The word graphs are searched for trade names or active substances and, subsequently, matching dosages. New concept edges are inserted if valid prescriptions are found in the word graph.

4.1 Detecting Medications in the Word Graph

The (multi-edge) word graphs are scanned, and the words associated with each edge are compared to the medications in the knowledge base. Figure 1 shows a word graph consisting of hypotheses generated by ASR, which is the input to our system. The dashed edges indicate the best path, while dotted lines are hypotheses which are not on the best path.

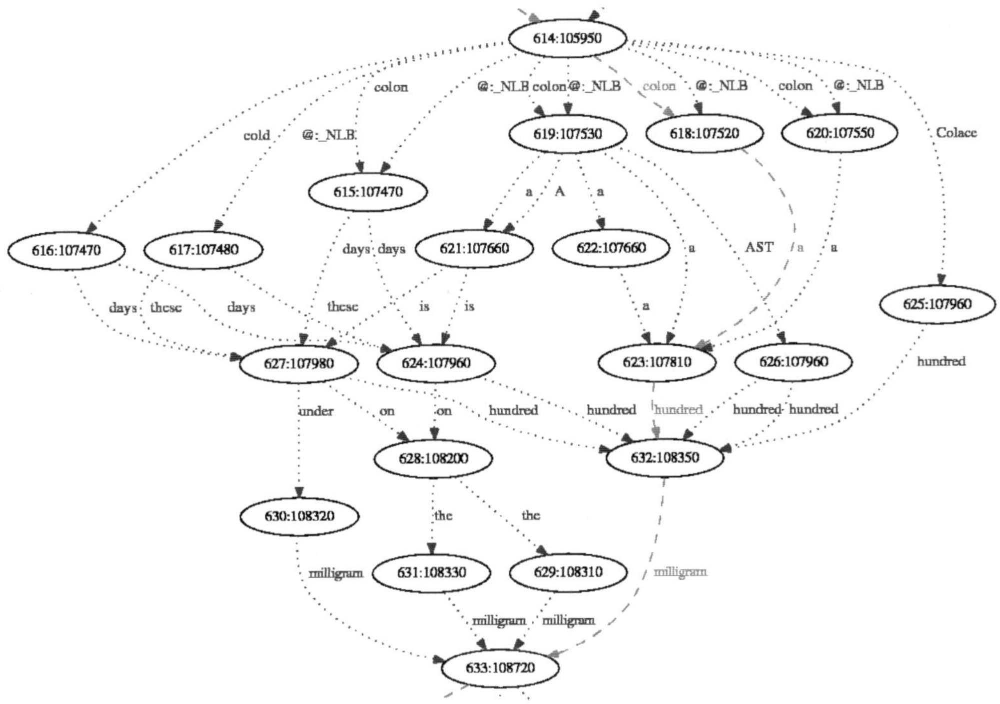

Figure 1: Sample word graph fragment

In case a match, i.e. a trade name or an active substance, is found, all edges succeeding the medication edge are searched for dosage values and dosage units. So far, we only examine the context to the right-hand side; in the data, we did not encounter any medications where the dosage occurred before the trade name or active substance. The following kinds of fillers between the trade name or active substance and the dosage are allowed: 'to' and 'of' as well as non-utterances such as *hesitation*, *noise* and *silence*; in the corpus, we did not encounter any other fillers.

4.2 Generation of Spoken Forms and Mapping

The medication found in the word graph is looked up in RxNorm, and all possible spoken forms of valid dosage values and dosage units for this medication are generated. Spoken forms for the medication names consist of the trade names and the active substances. Variation in the pronunciation of the trade names or active substances is handled by the ASR recognition lexicon. For generating spoken forms of the dosage values, finite-state tools were used. For dosage units, we wrote a small grammar. Looking at two examples, the medication *Synthroid*® and *Colace*® (the latter appears in the word graphs in Figure 2 and Figure 1), the spoken forms shown in Table 1 are generated. Each box contains the alternative spoken variants. *Synthroid*® contains the active substance Thyroxine and *Colace*® contains the active substance Docusate; users may either refer to the trade name or the active substance, so both possibilities are generated for each medication and dosage. RxNorm does not contain the dosage unit 'mcg' (micrograms), which occurred in the reports. Therefore, microgram dosage values were converted to milligrams. Since both 'miligram(s)' and 'microgram(s)' may occur for *Synthroid*®, dosage values for both dosage units are generated. Although strictly, 'twenty five' and 'twenty-five' are identical spoken forms, both versions may appear in the word graph and thus are provided by our system.

Sometimes, a medication may contain several active substances, e.g. *Hyzaar*®, a medication against high blood pressure:

Hydrochlorothiazide 12.5 MG / Losartan 50 MG Oral Tablet [Hyzaar]

trade name/ active substance	dosage value	dosage unit
'Synthroid' 'Thyroxine'	'zero point zero two five' 'zero point O two five' 'O point zero two five' 'O point O two five' 'point zero two five' 'point O two five'	'milligram' 'milligrams'
	'twenty five' 'twenty-five' 'two five'	'microgram' 'micrograms'
'Colace' 'Docusate'	'one hundred' 'a hundred' 'hundred'	'miligram' 'miligrams'

Table 1: Generated spoken forms found in the word graph

In these cases, the generation of possible spoken forms also includes different permutations of substances, as well as a spoken forms containing the dosage unit either only at the end or after each value if the dosage unit is identical.

4.3 Inserting Concept Edges

The sequences of words which constitute the word graph are compared to the spoken forms generated for the RxNorm knowledge base. The active substances or trade names serve as a starting point: in case a trade name is found in the word graph, the spoken forms for dosages of all active substances are generated in all permutations. If an active substance is found in the word graph, only the spoken forms for the substance dosage are searched in the word graph.

A new concept edge is inserted into the word graph for each path matching one of the generated spoken forms of the medications data base. The inserted concept edges span from the first matching node to the last matching node on the path. Figure 2 shows the word graph from Figure 1 with an inserted concept edge (in bold). For each inserted concept edge, new concept-edge attributes are assigned containing the IDs of the original edges as children, their added scores plus an additional concept score and the sequence of words. Since no large-scale experiments have yet been carried out, so far the concept score which is added to the individual scores of the children is an arbitrary number which improves the score of the medication subpath in constrast to paths which do not contain valid medication information. If several competing medication paths are found, a concept edge is inserted for each path, and the concept edges can be ranked according to their acoustic and language-model scores.

5 Evaluation

In the first step, we examined a report sample in order to determine if there are cases where a valid prescription is recognised although the physician did not mention a prescription. We did not encounter this phenomenon in our report corpus.

We then applied our method to a sample of 924 word graphs. In this sample,

- 481 valid dosages could be found, although
- only 325 of these were on the best path.

With our approach, for the 156 prescriptions (32%) which were not on the best path, alternatives could be reconstructed from the word graph. Based on the inserted concept edges, the best path can be rescored.

In order to measure recall, i.e. how many of all existing prescriptions in the reports can be detected with our knowledge base, we manually checked a sample of 132 reports (containing manual transcriptions and ASR results). In this sample, 85 errors concerning medications and/or prescriptions occurred. For 19 of the 85 errors, the correct result was contained in the word graph. For 8 errors, it could be reconstructed. So about 9% of the errors concerning medications can be corrected in our sample. For the cases where the prescription could not be reconstructed although it was contained in the word graph, an analysis of the errors is shown in Table 2.

Since new medications are constantly being released, and trade names change frequently, mismatches may be due to the fact that our version of RxNorm was from a more recent point in time than the report corpus. We assume that under real-world conditions, both RxNorm and the medications prescribed by physicians reflect the current situation.

Some problems concerning medication names and dosage units were caused by missing spoken forms containing abbreviations, e.g. of dosage units (*mg* vs. *mg/ml*) or names (*Lantus* vs. *Lantus insulin*). Here, the coverage needs to be improved.

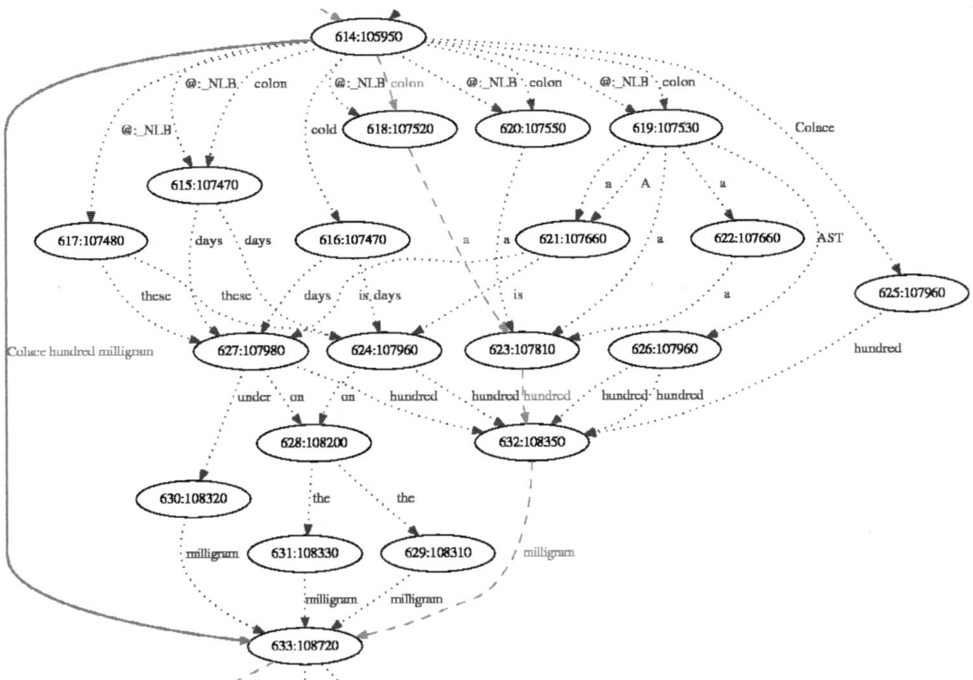

Figure 2: Sample word graph fragment with inserted concept node (left)

Table 2: Error types found in manual evaluation

type of error	#	example	
		Word Graph	**RxNorm**
differences in medication names between the knowledge base and the word graph	3	*Cardizem CD 120 mg*	*Cardizem 120 mg*
differences in dosage values between the knowledge base and the word graph	4	*Tapazole 60 mg*	*Tapazole 10 mg*
differences in dosage units between the knowledge base and the word graph	4	*Epogen 20000 units*	*Epogen 20000 ml*

There are also cases where two medications appear in the word graph, and both had the valid prescription strength, therefore the system was not able to determine the correct medication.

6 Conclusion

In this paper, we present an attempt to reduce the number of speech-recognition errors concerning prescriptions of medications based on a domain-specific knowledge base. Our approach uses word graphs as input and creates new versions of the word graph with inserted concept edges if more plausible prescriptions are found. The concept edges can be used for rescoring the best path. An evaluation showed that 32% of prescriptions found in the word graphs were not on the best path but could be reconstructed. The manual evaluation of 132 reports shows that our method covers 42% of the prescriptions which are actually spoken during the dictation.

At present, we have only investigated the reduction of medication misrecognitions in our evaluation. In a larger evaluation, we will determine the actual impact of our method on the word-error rate of medical reports. Furthermore, we are working on integrating additional available knowledge sources so that the plausibility of prescriptions can also be assessed from a broader medical point of view, e.g. in case two subsequent prescriptions are encountered in the word graph which are incompatible due to drug interactions. As a next step, the system can be extended to compare the prescriptions with the patient record, e.g. if a patient has medication allergies. So far, our simple solution integrating only available, constantly updated knowledge about medications has already turned out to be a good starting point for rescoring word graphs based on domain knowledge.

Acknowledgments

The work presented here has been carried out in the context of the Austrian KNet competence network COAST. We gratefully acknowledge funding by the Austrian Federal Ministry of Economics and Labour, and ZIT Zentrum fuer Innovation und Technologie, Vienna. The Austrian Research Institute for Artificial Intelligence is supported by the Austrian Federal Ministry for Transport, Innovation, and Technology and by the Austrian Federal Ministry for Science and Research. The authors would like to thank the anonymous reviewers for their helpful comments.

References

A. Alapetite, A., H.B. Andersen, H.B. and M. Hertzumb. Acceptance of speech recognition by physicians: A survey of expectations, experiences, and social influence. *International Journal of Human-Computer Studies* **67**(1) (2009) 36–49

D. Bühler, W. Minker and A. Elciyanti. Using language modelling to integrate speech recognition with a flat semantic analysis. In: *6th SIGdial Workshop on Discourse and Dialogue*, Lisbon, Portugal (September 2005) http://www.sigdial.org/workshops/workshop6/proceedings/pdf/86-paper.pdf.

S. Gold, N. Elhadad, X. Zhu, J.J. Cimino, G. Hripcsak. Extracting Structured Medication Event Information from Discharge Summaries. In: *Proceedings of the AMIA 2008 Symposium.*

I. Gurevych and R. Porzel. Using knowledge-based scores for identifying best speech recognition hypothesis. In: *Proceedings of ISCA Tutorial and Research Workshop on Error Handling in Spoken Dialogue Systems*, Chateau-d'Oex-Vaud, Switzerland (2003) 77–81 http://proffs.tk.informatik.tu-darmstadt.de/TK/abstracts.php3?lang=en&bibtex=1&-paperID=431.

R. Porzel, I. Gurevych and C. Müller. *Ontology-based contextual coherence scoring*. Technical report, European Media Laboratory, Heidelberg, Germany (2003) http://citeseer.ist.psu.edu/649012.html.

S.R. Halgrim, F. Xia, I. Solti, E. Cadag and O. Uzuner. Statistical Extraction of Medication Information from Clinical Records. In: *Proc. of AMIA Summit on Translational Bioinformatics*, San Francisco, CA, March 10-12, 2010.

D.A. Lindberg, B.L. Humphreys and A.T. McCray. The unified medical language system. *Methods of Information in Medicine* **32**(4) (August 1993) 281–291 http://www.nlm.nih.gov/research/umls/.

S. Liu, W. Ma, R. Moore, V. Ganesan and S. Nelson. Rxnorm: Prescription for electronic drug information exchange. *IT Professional* **7**(5) (September/October 2005) 17–23

K. Macherey, O. Bender and H. Ney. Multi-level error handling for tree based dialogue course management. In: *Proceedings of ISCA Tutorial and Research Workshop on Error Handling in Spoken Dialogue Systems*, Chateau-d'Oex-Vaud, Switzerland (2003) 123–128, http://www-i6.informatik.rwth-aachen.de/~bender/papers/isca_tutorial_2003.pdf.

M. Oerder and H. Ney. Word graphs: An efficient interface between continuous speech recognition and language understanding. In: *Proc. IEEE ICASSP'93*. Volume 2. 119–122.

J. Patrick and M. Li. A Cascade Approach to Extracting Medication Events. In: *Proc. Australasian Language Technology Workshop (ALTA) 2009.*

R. Sarikaya, Y. Gao and M. Picheny. Word level confidence measurement using semantic features. In: *Proc. of IEEE ICASSP2003*. Volume 1. (April 2003) 604–607.

S. Seneff and J. Polifroni. Dialogue Management in the MERCURY Flight Reservation System. In: *Satellite Dialogue Workshop, ANLP-NAACL*, Seattle (April 2000).

K.D. Voll. *A Methodology of Error Detection: Improving Speech Recognition in Radiology*. PhD thesis, Simon Fraser University (2006) http://ir.lib.sfu.ca/handle/1892/2734.

K. Wang, Y.Y. Wang and A. Acero. Use and acquisition of semantic language model. In: *HLT-NAACL*. (2004) http://www.aclweb.org/anthology-new/N/N04/N04-3011.pdf.

R. Zhang and A.I. Rudnicky. Word level confidence annotation using combinations of features. In: *Proceedings of Eurospeech*. (2001) http://www.speech.cs.cmu.edu/Communicator/papers/-RecoConf2001.pdf.

Assessment of Utility in Web Mining for the Domain of Public Health

Peter von Etter, Silja Huttunen, Arto Vihavainen, Matti Vuorinen and **Roman Yangarber**
Department of Computer Science
University of Helsinki, Finland
First.Last@cs.helsinki.fi

Abstract

This paper presents ongoing work on application of Information Extraction (IE) technology to domain of Public Health, in a real-world scenario. A central issue in IE is the *quality* of the results. We present two novel points. First, we distinguish the criteria for quality: the objective criteria that measure correctness of the system's analysis in traditional terms (F-measure, recall and precision), and, on the other hand, subjective criteria that measure the *utility* of the results to the end-user.

Second, to obtain measures of utility, we build an environment that allows users to interact with the system by rating the analyzed content. We then build and compare several classifiers that learn from the user's responses to predict the relevance scores for new events. We conduct experiments with learning to predict relevance, and discuss the results and their implications for text mining in the domain of Public Health.

1 Introduction

We describe an on-going project for text mining in the domain of Public Health. The aim of the project is to build a system for providing decision support to Public Health (PH) professionals and officials, in the task of Epidemic Surveillance.

Epidemic surveillance may be sub-divided into *indicator-based* vs. *event-based* surveillance, (Hartley et al., 2010). Whereas the former is based on structured, quantitative data, which is collected, e.g., from national or international clinical laboratories or databases, and is of reliable quality, the latter is much more noisy, and relies on "alert and rumour scanning", particularly from *open-source media*, such as on-line news sites. While the latter kind of information sources are less reliable overall, they nonetheless constitute a crucial channel of information in PH. This is because the media are extremely adept at picking up isolated cases and weak signals—which may be indicative of emergence of important events, such as an incipient epidemic or critical change in a public-health situation—and in many cases they can do so much more swiftly than official channels. National and supra-national (e.g., European-level) Health Authorities require timely information about threats posed to the public by emerging infectious diseases and epidemics. Therefore, these Agencies rely on media-monitoring as a matter of routine, on a continual basis as part of their day-to-day operations.

The system described in this paper, PULS, is designed to support Epidemic Surveillance by monitoring open-source media for reports about events of potential significance to Public Health (Yangarber and Steinberger, 2009). We focus in this paper on news articles mentioning incidents of infectious diseases. The system does not make decisions, but provides decision support, by filtering massive volumes of information and trying to identify those cases that should be brought to the attention of *epidemic intelligence officers* (EIO)—public health specialists engaged in epidemic surveillance.

This is an inter-disciplinary effort. The system builds on methods from text mining and computational linguistics to identify the items of potential interest (Grishman et al., 2003). The EIOs, on the other hand, are medical professionals, and are generally not trained in computational methods. Therefore the tools that they use must be intuitive and must

Proceedings of the NAACL HLT 2010 Second Louhi Workshop on Text and Data Mining of Health Documents, pages 29–37,
Los Angeles, California, June 2010.

not overwhelm the user with volume or complexity.

A convenient baseline for comparison is keyword-based search, as provided by search engines and news aggregators. Systems that rely on keyword-matching to find articles related to infectious threats and epidemics quickly overwhelm the user with a vast amount of news items, much of which is noise.

We have tuned PULS, the "Pattern-based Understanding and Learning System," to support Epidemic Surveillance in several phases. PULS is a collaborative effort with MedISys, a system for gathering epidemic intelligence built by the European Commission (EC) at the Joint Research Centre (JRC) in Ispra, Italy. First, MedISys finds news articles from thousands of on-line sources around the world, identifies articles potentially relevant to Epidemic Surveillance, using a broad keyword-based Web search, and sends them via an RSS feed to PULS on a continual basis. Second, PULS employs "fact-finding" technology, *Information Extraction* (IE), to determine exactly what happened in each article: who was affected by what disease/condition, where and when—creating a structured record that is stored in the database. Articles that do not trigger creation of a database record are discarded. A third component then determines the *relevance* of the selected articles—and cases that they describe—to the domain of Public Health, specifically to Epidemic Surveillance.

Traditionally in IE research, performance has been measured in terms of formal *correctness*—how accurately the system is able to analyze the article (Hirschman, 1998). In this paper we argue the need for other measures of performance for text mining, using as a case study our application of Web mining to the domain of Public Health. In the next section, we lay down criteria for judging *quality*, and present the approach taken in our system. Section 3 outlines the organisation of the system, and Section 4 presents in detail our experiments with automatic assignment of relevance scores. In the final section we discuss the results and outline next steps.

2 Criteria for quality

In this section we take a critical view at traditional measures of *quality*, in text analysis in general, and IE in particular. What defines quality most appropriately for our application, and how should we measure quality? We propose the following taxonomy of quality in our context:

- Objective: system's perspective
  - Correctness
  - Confidence
- Subjective: user's perspective
  - Utility or relevance
  - Reliability

At the top level, we distinguish *objective* vs. *subjective* measures. Most IE research has focused on correctness over the last two decades, e.g., in the MUC and ACE initiatives (Hirschman, 1998; ACE, 2004). Correctness is a measure of how accurately the system extracts the semantics from an article of text, in terms of matching the system's answers to a set of answers pre-defined by human annotators. In our context, a set of articles is annotated with a "gold-standard" set of database records, each record containing fields like: the name of the disease/infectious agent, the location/country of the incident, the date of the incident, the number of victims, whether they are human or animal, whether they survived, etc. Then the system's response can be compared to the gold standard and correctness can be computed in terms of recall and precision, F-measure, accuracy, etc.—counting how many of the fields in each record were correctly extracted. This approach to quality is similar to the approach taken in other areas of computational linguistics: how many structures in the text were correctly identified, how many were missed, and how many spurious structures were introduced.

Confidence has been studied as well, to estimate the probability of the correctness of the system's answer, e.g., in (Culotta and McCallum, 2004). Our system computes *confidence* using discourse-level cues, (Huttunen et al., 2002): e.g., confidence decreases as the distance between event trigger and event attributes increases—the sentence that mentions that someone has fallen ill or died is far from the mention of the disease. Confidence also depends on uniqueness of attributes—e.g., if a document mentions only one country, the system has more confidence that an event referring to this country is correct.

On the subjective side, utility, or relevance, asks how *useful* the result is to the user. There are several points to note. First, it is clearly a highly subjective measure, not easy to capture in exact terms. Second, it is "orthogonal" to correctness in the sense that from the user's perspective utility matters *irrespective* of correctness. For example, an extracted case can be 100% correct, yet have very low utility to the user, (for the task of epidemic surveillance)—a perfectly extracted event that happened too long ago would not matter in the current context. Conversely, every slot in the record may be extracted erroneously, and yet the event may be of great importance and *value* to the user. We focus specifically on relevance vs. correctness.

Given the current performance "ceilings" of 70-80% F-measure in state-of-the-art IE, what does correctness of $x\%$ mean in practice? It likely means that if $x > y$ then a system achieving F-measure x is better to have than one achieving y. But what does it say about *utility*? In the best case, correctness may be correlated with utility, in the worst case it is independent of utility (e.g., if the system happens to achieve high correctness on events from the past, which have low relevance). Since we are targeting a specific user base, the user's perspective must be taken into account when estimating quality, not (only) the system's perspective. This implies the need for *automatic assignment of relevance* scores to analyzed events or documents.

Finally, *reliability* measures whether the reported event is "true". The relevance of extracted fact may be high, but is it credible? Can the information be trusted? We list this criterion for quality for completeness, since it is the ultimate goal of any surveillance process. However, answering this requires a great deal of knowledge external to the system, that can only be obtained by the human user through a detailed down-stream verification process. The system may provide some support for determining reliability, e.g., by tracking the performance of different information sources over time, since the reliability of the facts extracted from an article is related to the reliability of the source. It may be possible to classify Web-based sources according to their credibility; some sources may habitually withhold information (for fear of impact to tourism, trade, etc.); other sites may try to attract readership by exaggerated claims (e.g., tabloids). On the other hand, clearly disreputable sites may carry true information. This measure of quality is beyond the scope of this paper.

3 The System: Background

PULS, the *Pattern-based Understanding and Learning System*, is developed at the University of Helsinki to extract factual information from plain text. PULS has been adapted to analyse texts for Epidemic Surveillance.[1]

The components of PULS have been described in detail previously, (Yangarber and Steinberger, 2009; Steinberger et al., 2008; Yangarber et al., 2007). In several respects, it is similar to other existing systems for automated epidemic surveillance, viz., BioCaster (Doan et al., 2008), MedISys and PULS (Yangarber and Steinberger, 2009), HealthMap (Freifeld et al., 2008), and others (Linge et al., 2009).

PULS relies on EC-JRC's MedISys for IR (information retrieval)—MedISys performs a broad Web search, using a set of boolean keyword-based queries, (Steinberger et al., 2008). The result is a continuous stream of potentially relevant documents, updated every few minutes. Second, an IE component, (Grishman et al., 2003; Yangarber and Steinberger, 2009), analyzes each retrieved document, to try to find events of potential relevance to Public Health. The system stores the structured information about every detected event into a database. The IE component uses a large set of linguistic patterns, which in turn depend on a large-scale public health ontology, similar to MeSH,[2] that contains concepts for diseases and infectious agents, infectious vectors and animals, medical drugs, and geographic locations.

From each article, PULS's pattern matching engine tries to extract a set of incidents, or "facts"—detailed information related to instances of disease outbreak. An incident is described by a set of fields, or attributes: location and country of the incident, disease name, the date of the incident, information about the victims—their type (people, animals, etc.),

[1] puls.cs.helsinki.fi/medical

[2] www.nlm.nih.gov/mesh

number, whether they survived or died, etc.

The result of IE is a populated database of extracted items, that can be browsed and searched by any attribute, according to the user's interests. It is crucial to note that the notion of a user's *focus* or interest is **not** the same as the notion of relevance, introduced above. We take the view that the notion of relevance is shared among the entire PH community: an event is either relevant to PH or it is not. Note also, that this view is upheld by several classic, *human-moderated* PH surveillance systems, such as ProMED-Mail[3] or Canadian GPHIN. User's interest is individual, e.g., a user may have specific geographic, or medical focus (e.g., only viral or tropical illnesses), and given the structured database, s/he can filter the content according to specific criteria. But that is independent of the *shared* notion of relevance to PH. User focus can be exploited for targeted recommendation, using techniques such as collaborative filtering; at present, this is beyond the scope of our work.

The crawler and IE components have been in operation and under refinement for some time. We next build a classifier to assign relevance scores to each extracted event and matched document.

4 Experimental Setup

We now present the work on automatic classification of relevance scores. In collaboration with the end-users, we defined guidelines for judging relevance on a 6-point scale, summarized in Table 1.

Criteria	*Score*
New information, highly relevant	5
Important updates, on-going developments	4
Review of current events, potential risk of disease	3
Historical/non-current events Background information	2
Non-specific, non-factive events, secondary topics, scientific studies hypothetical risk	1
Unrelated to PH	0

Table 1: Guidelines for relevance scores in medical news

[3] www.promedmail.org

Note, the separation between the "high-relevance" scores, 4 and 5, vs. the rest; this split is addressed in detail in Section 4.3.

4.1 Discourse features

It is clear that these guidelines are highly subjective, and cannot be encoded by rules directly. In order to model the relevance judgements, we extracted features—the *discourse features*—from the document that are indicative of, or mappable to, the relevance scores. Discourse features try to capture higher-order information, including complex and longer-range inter-dependencies and clues, involving the physical layout of the document, and deeper semantic and conceptual information found in the document. Some examples of discourse features are:

- *Relative-position*, which is represented by a number from zero to 1 indicating the proportion of the document one needs to read to reach the event text;
- *Disease-in-header* is a binary value that indicates whether the disease is mentioned in the headline or the first two sentences;
- *Disease-to-trigger-distance* indicates how far the disease is from the trigger sentence (same as for confidence computation);
- *Recency* is the number of days between the reported occurrence of the event and the publication date;

We compiled over two dozen discourse-level features. It is clear that the discourse features do not determine the relevance scores, but provide weak indicators of relevance, so that probabilistic classification is appropriate. For example, a higher relative position of an event probably indicates lower relevance, but there are often *news summary* articles that gather many unrelated news together, and may contain very important items anywhere in the article.[4] A feature such as *Victim-named*, stating whether the victim's name is mentioned, often indicates lower-relevance events (obituaries, stories

[4] Due to space limitations, we do not provide a detailed list of the discourse features.

about public personalities, etc.). However, sometimes news articles about disease outbreaks deliberately personify the victims, to give the reader a sense of their background, lifestyle, to speculate about the victims' common circumstances.

We describe two classifiers we have built for relevance. A Naive Bayes classifier (NB) was used as the baseline. We then tried to obtain improved performance with Support Vector Machines (SVM).

4.2 Data

The dataset is the database of facts extracted by the system. The system pre-assigns relevance to each event, and users have the option to accept or correct the system's relevance score, through the User Interface, which also allows the users to correct erroneous fills, e.g., if a country, disease name, etc., was extracted incorrectly by the system.

Along with the users, members of the development team also evaluated a sample of the extracted events, and corrected relevance and erroneous fills. The developers are computer scientists and linguists, whereas the users are medics, and because they interpreted the guidelines differently this had an impact on the results, described in Tables 2 and 5.

"Cleaned data": PULS's user interface also permits users to **correct** incorrect fills in the events (in the two rightmost columns in the tables). This allowed us to obtain two parallel sets of examples with relevance labels: the raw examples, as they were automatically extracted by the system, and the "cleaned" examples, after users/developer corrections. The raw set is more noisy, since it contains errors introduced by the system. We used the cleaned examples to train our classifiers, and tested them on both the cleaned set and the raw set. Testing against the cleaned set gives an "idealized" performance, (as if the IE system made no errors in analysis). True performance is expected be closer to testing on the raw set.

In total, there were just under 1000 examples labeled by the users and the developers (some examples were labeled by both, since the system allows multiple users to attach different relevance judgements to the same example. Most of the time users agreed on the relevance judgements, but non-developers were less likely to clean examples.)

4.3 Naive Bayes classifier

Initially, we planned to perform regression to the complete [0–5] relevance scale. However, this proved problematic, since the amount of labeled data was not sufficient to cover the continuum between highly relevant and not-so-relevant items. We therefore decided instead to build a *binary* classifier. This decision is also justified in the context of our system's user interface, which provides the users with two views:

- the *Front Page View* contains only high-relevance items (rated 4 or 5), in case the user wants to see only the most urgent items first;
- the *Complete View* shows the user all extracted items, irrespective of relevance. (The user can always filter the database by relevance value.)

Thus, the relevance score is also used to guide a *binary* decision: whether to present a given event/article to the user on the Front-Page View. The NB classifier using the entire set of discourse features did not perform well, because the discourse features we have implemented are inherently not independent, which affects the performance of NB.

To try to reduce the mutual dependence among the features, we added a simple, *greedy* feature-selection phase during training. Feature selection starts by training a classifier on the full set of features, using leave-one-out (LOO) cross-validation to estimate the classifier's performance. In the next phase, the algorithm in turn excludes the features one by one, and runs the LOO cross-validation again, once with each feature excluded. The feature whose exclusion gives rise to the biggest increase in performance is dropped out, and the selection step is repeated with the reduced set of features. We continue to drop features until performance does not increase for several iterations; in our experiments, we used three steps beyond the top performance. We then back up to the step that yielded peak performance. The resulting subset of features is used to train the final NB classifier.

The NB classifier is implemented in R Language.

Because relevance prediction is difficult for all events, we also tried to predict the relevance of an article, making the simplifying assumption that the article is only as relevant as the *first* event found in the article.[5] The results are presented in Table 2. The rows labeled *Dev only* refer to the data sets labeled by developers, and *Users only* to sets labeled by (non-developer) users.

	Testing on		Number examples	
	Clean	Raw	Clean	Raw
	Event-level			
Dev only	76.96	**76.66**	560	510
All	72.19	73.34	863	799
Users only	70.38	66.53	303	289
	Document-level			
Dev only	80.41	**79.00**	291	281
All	73.94	72.45	545	530
Users only	65.82	67.09	238	232

Table 2: Naive Bayes prediction accuracy

The event-level classification is shown in the top portion of the table. Throughout, as expected, testing on the cleaned data usually gives slightly better (more idealized) performance estimates than testing on the raw. Also, as expected, testing on the first-only events (document-level) gives slightly better performance, since it's a simpler problem—although there is less data to train/test on.

It is important to observe that using data labeled by developers gives significantly higher performance. This is because coercing the users to follow the guidelines strictly is not possible, and they deviate from the rules that they themselves helped articulate. The rows labeled "all" show performance when all combined available data was used—labeled by both the developers and the users.

This performance is quite good for a baseline.[6] The confusion matrices—for the developer-only event-level raw data set—show the distribution of true/false positives/negatives.

4.4 SVM Classifier

For comparison, we built two additional classifiers using the SVMLight Toolkit.[7] We first used a linear kernel as a baseline, and used a RBF kernel, which is potentially more expressive. The conditions for testing the SVM classifiers were same as the ones for the NB classifiers, and same datasets were used as for the NB.

	True Labels	
Predicted labels	4-5	0-3
High-relevance 4-5	125	77
Low-relevance 0-3	42	266

Table 3: NB confusion matrix

As SVM with the RBF kernel can use non-linear separating hyperplanes in the original feature space by using the kernel trick (Aizerman et al., 1964), we aimed to test whether it would provide an improvement over the linear kernel. (For more detailed discussions of SVM and different kernel functions for text classification, cf., for example, (Joachims, 1998).)

To regularize the input for SVM, all feature values were normalized to lie between 0 and 1 (for continuous-valued features), and set to 0 or 1 for binary features. Table 4 describes the accuracy achieved with the linear kernel. Experiments labeled *All discourse features* use the complete set of discourse features (over 20 features). Rows labeled *Selected discourse features* show results from training with exactly same features as resulted from the feature selection phase of NB.

	Event-level		*Document-level*	
	Clean	Raw	Clean	Raw
	All discourse features			
Dev only	75.33	**77.17**	76.87	76.56
All	71.60	72.26	70.51	69.96
	Selected discourse features only			
Dev only	76.07	**77.95**	77.94	77.62
All	71.40	72.14	69.75	69.37

Table 4: SVM prediction accuracy using linear kernel

The difference when training with selected discourse features and all discourse features is not large, since SVM is able to distinguish between relevant and non-relevant features fairly well. The results from SVM using linear kernel appear compa-

[5] A manual check confirmed that there were *no* instances where the first event in an article had lower relevance than a subsequent event.

[6] Consider for comparison, that the *correctness* on a manually constructed, non-hidden set of articles used for system development, is under 75% F-measure.

[7] http://svmlight.joachims.org/

rable with the results from the NB.

In addition to using the discourse features, we also tried using *lexical features*. The lexical features for a given example—extracted event—is simply the bag of words from the sentence containing the event, plus the two surrounding sentences. To reduce data sparsity, the sentences are pre-processed by a lemmatizer, and passed through a named entity (NE) recognizer (Grishman et al., 2003), which replaces persons, organizations, locations and disease names with a special token indicating the NE's class. "Stop-word" parts of speech were dropped—prepositions, conjunctions, and articles.

	Event-level		*Document-level*	
	Clean	Raw	Clean	Raw
All discourse features				
Dev only	74.69	**75.37**	77.93	78.38
All	69.58	70.26	71.56	71.25
Selected discourse features only				
Dev only	77.51	**79.01**	79.19	79.04
All	72.02	72.84	72.59	72.30
Lexical features only				
Dev only	75.93	**76.37**	79.11	80.07
All	73.28	73.47	74.53	74.71
Lexical and selected discourse features				
Dev only	78.87	**79.24**	82.66	81.83
All	76.48	76.58	76.52	76.19

Table 5: SVM prediction accuracy using RBF kernel

The performance of SVM with the RBF kernel is strongly dependent on the values of SVM parameters C—the trade-off between training error and margin— and γ—the kernel width (Joachims, 1998). We tuned these parameters manually by checking a grid of values against a development dataset, and finding areas where the SVM performed well. These areas were then further investigated. After trying 40 combinations, we set C as 10000 and γ to 0.001 for subsequent evaluations. The results for SVM using RBF kernel are given in Table 5.

High accuracy of lexical features alone was somewhat surprising as lexical features consist only of the bag of words in the event-bearing sentence, plus the preceding and the following sentences. News articles often have various pieces of information related to the event scattered around the document. For example, the disease can appear only in the headline, the location/country in the middle of the document, and the event-bearing sentence in a third location, (Huttunen et al., 2002). Our lexical features, as presented here, are not capable of capturing such long-distance relationships.

The observed difference in performance on relevance prediction between the data sets labeled by developers vs. non-developer users, likely arises from the fact that developers follow the formal guidelines more strictly (being computer scientists). Rows labeled *all* show performance against data sets labeled by real users, who work in different PH organizations in several different countries, each group of users intuitively following their own, subjective guidelines, *despite* the common guidelines agreed-upon for this project. There may also be deviation within organizations. For example, certain doctors may find specific diseases or locations more interesting, giving events containing them a high relevance, thus injecting personal preference into document relevance.

5 Discussion and Conclusions

The SVM performs somewhat better than the Naive Bayes classifier, though there is still much to be explored and improved. One odd effect is that sometimes testing on the raw data gives slightly better results than testing on the clean data, though this is probably not significant, since the SVM classifier is still not finely tuned (and the data contain some noise). Using all discourse features performs slightly worse than using a reduced set of features—the same set of features that we obtained through greedy feature selection for NB.

Although the lexical features alone seem to do somewhat worse than the discourse features alone on event-level classification, we still see that the lexical features contain a great deal of information (which the NB cannot use). As expected, adding the discourse features improves performance over lexical features alone, since discourse features capture information about long-range dependencies that local lexical features do not.

In forming splits for cross-validation or LOO, we made sure not to split examples from the same document across the training and test sets. That is, for a given document, *all* events in it are either used for training or for testing, to avoid biasing the testing.

To summarize, the points addressed in this paper:

- We have presented a language-technology-based approach to a problem in Public Health, specifically the problem of event-based epidemic surveillance through monitoring on-line media.
- The user's perspective needs to be taken into account when estimating quality, not just the system's perspective. *Utility* to the user is at least as important as (if not more important than) correctness.
- We have presented an operational system that suggests articles potentially relevant to the user, and assigns relevance scores to each extracted event.
- For now, we assume the users share same notion of relevance of an event to Public Health.
- We have presented experiments and an initial evaluation of assignment of relevance scores.
- Experiments indicate that relevance appears to be a *tractable* measure of quality, at least in principle. Marking document-level relevance—only for the first event in the document—appears to be easier. However, making real users follow strict guidelines is difficult in practice.

On-going work includes refining the classification approaches, especially, using Bayesian networks, regression, using transductive SVMs to leverage unlabeled data, and exploring collaborative filtering to address users' individual interests.

Acknowledgments

This research was supported in part by: the Technology Development Agency of Finland (TEKES), through the ContentFactory Project, and by the Academy of Finland's National Centre of Excellence "Algorithmic Data Analysis (ALGODAN)."

References

ACE. 2004. Automatic content extraction.

M. A. Aizerman, E. A. Braverman, and L. Rozonoer. 1964. Theoretical foundations of the potential function method in pattern recognition learning. In *Automation and Remote Control*, volume 25, pages 821–837.

Aron Culotta and Andrew McCallum. 2004. Confidence estimation for information extraction. In *Proceedings of Human Language Technology Conference and North American Chapter of the Association for Computational Linguistics*.

Son Doan, Quoc Hung-Ngo, Ai Kawazoe, and Nigel Collier. 2008. Global Health Monitor—a web-based system for detecting and mapping infectious diseases. In *Proceedings of the International Joint Conference on Natural Language Processing (IJCNLP)*.

C.C. Freifeld, K.D. Mandl, B.Y. Reis, and J.S. Brownstein. 2008. HealthMap: Global infectious disease monitoring through automated classification and visualization of internet media reports. *Journal of American Medical Informatics Association*, 15:150–157.

Ralph Grishman, Silja Huttunen, and Roman Yangarber. 2003. Information extraction for enhanced access to disease outbreak reports. *Journal of Biomedical Informatics*, **35**(4):236–246.

David Hartley, Noele Nelson, Ronald Walters, Ray Arthur, Roman Yangarber, Larry Madoff, Jens Linge, Abla Mawudeku, Nigel Collier, John Brownstein, Germain Thinus, and Nigel Lightfoot. 2010. The landscape of international event-based biosurveillance. *Emerging Health Threats Journal*, 3(e3).

Lynette Hirschman. 1998. Language understanding evaluations: Lessons learned from muc and atis. In *Proceedings of the First International Conference on Language Resources and Evaluation (LREC)*, pages 117–122, Granada, Spain, May.

Silja Huttunen, Roman Yangarber, and Ralph Grishman. 2002. Complexity of event structure in information extraction. In *Proceedings of the 19th International Conference on Computational Linguistics (COLING 2002)*, Taipei, August.

Thorsten Joachims. 1998. Text categorization with suport vector machines: Learning with many relevant features. In *ECML: European Conference on Machine Learning*, pages 137–142.

J.P. Linge, R. Steinberger, T.P. Weber, R. Yangarber, E. van der Goot, D.H. Al Khudhairy, and N.I. Stilianakis. 2009. Internet surveillance systems for early alerting of health threats. *Eurosurveillance Journal*, 14(13).

Ralf Steinberger, Flavio Fuart, Erik van der Goot, Clive Best, Peter von Etter, and Roman Yangarber. 2008.

Text mining from the web for medical intelligence. In Domenico Perrotta, Jakub Piskorski, Franoise Soulié-Fogelman, and Ralf Steinberger, editors, *Mining Massive Data Sets for Security*. OIS Press, Amsterdam, the Netherlands.

Roman Yangarber and Ralf Steinberger. 2009. Automatic epidemiological surveillance from on-line news in MedISys and PULS. In *Proceedings of IMED-2009: International Meeting on Emerging Diseases and Surveillance*, Vienna, Austria.

Roman Yangarber, Clive Best, Peter von Etter, Flavio Fuart, David Horby, and Ralf Steinberger. 2007. Combining information about epidemic threats from multiple sources. In *Proceedings of the MMIES Workshop, International Conference on Recent Advances in Natural Language Processing (RANLP 2007)*, Borovets, Bulgaria, September.

Reliability and Type of Consumer Health Documents on the World Wide Web: an Annotation Study

Melanie J. Martin
California State University, Stanislaus
One University Circle
Turlock, CA 95382
mmartin@cs.csustan.edu

Abstract

In this paper we present a detailed scheme for annotating medical web pages designed for health care consumers. The annotation is along two axes: first, by reliability or the extent to which the medical information on the page can be trusted, second, by the type of page (patient leaflet, commercial, link, medical article, testimonial, or support). We analyze inter-rater agreement among three judges for each category. Inter-rater agreement was moderate (0.77 accuracy, 0.62 F-measure, 0.49 Kappa) on the reliability axis and good (0.81 accuracy, 0.72 F-measure, 0.73 Kappa) along the type axis.

1 Introduction

With the explosive growth of the World Wide Web has come, not just an explosion of information, but also the explosion of false, misleading and unsupported information. At the same time, the web is increasingly used for tasks where information quality and reliability are vital, from legal and medical research by both professionals and lay people, to fact checking by journalists and research by government policy makers.

In particular, there has been a proliferation of web pages in the medical domain for health care consumers. At the first sign of illness or injury more and more people go to the web before consulting medical professionals. The quality and reliability of the information on consumer medical web pages has been of concern for some time to medical professionals and policy makers. (For example see Eysenbach et al., 2002, Impicciatore et al., 1997.)

Our goal is to create a system that can automatically measure the reliability of web pages in the medical domain (Martin, 2004). More specifically, given a web page resulting from a user query on a medical topic, we would like to automatically provide an estimate of the extent to which the information on the page can be trusted. In order to make use of supervised natural language processing and machine learning algorithms to create such a system, and to ultimately evaluate the performance of the system, it is necessary to have human annotated data.

It is important to note the varied uses of the term "reliability" in the computer and information sciences. In the current context we use it to refer to an intrinsic property of a web page: essentially the trustworthiness of the information it contains. This sense of reliability is distinct from its meaning in measurement theory as an indicator of repeatability. It also excludes measures such as credibility that are based on user beliefs or understanding.

In this paper we report results of an annotation study of medical web pages designed for health care consumers. Three humans annotated a corpus of web pages along two axes. The first axis is the reliability of the information contained in the page. The second axis is the type, or kind, of page. Intercoder agreement was moderate (0.77 accuracy, 0.62 F-measure, 0.49 Kappa) on the reliability axis and good (0.81 accuracy, 0.72 F-measure, 0.73 Kappa) along the type axis.

In our materials and methods section we discuss the data, definitions, annotation study and the results. We follow with a discussion section and a conclusion.

Proceedings of the NAACL HLT 2010 Second Louhi Workshop on Text and Data Mining of Health Documents, pages 38–45,
Los Angeles, California, June 2010.

2 Materials and Methods

In this section we will discuss the data and definitions for the annotation task. We also describe the annotation study and the testing and analysis.

2.1 Data

The data to be annotated consists of two corpora of web pages created by the author: IBS70 and MMED100. The MMED100 corpus is a subset of a larger corpus (MMED1000). Both corpora are described below.

2.1.1 IBS70 Corpus

The IBS70 corpus was created as an exploratory corpus for use in system development. It was originally the top 50 Google hits for "irritable bowel syndrome" downloaded automatically though the Google API on July 1, 2004. The query was chosen to provide a range of quality and types of pages which one would expect to see more generally in the medical domain on the web: patient information from both traditional and alternative sources, support groups, medical articles, commercial pages from drug companies and quacks. During system development we determined that it would be useful to have additional pages at both ends of the reliability spectrum, possibly to use as seeds for clustering.

On September 15, 2004, twenty documents were added to the corpus to create the IBS70. Ten highly reliable documents were added based on web searches to find documents judged as meeting the standards of Evidence Based Medicine. Ten documents judged unreliable were added by taking the first ten relevant “Sponsored Links” resulting from a Google search on “irritable bowel syndrome”. There are two important things to note about this process: first, the high quality pages added were disproportionately from the U.K.; second, the low quality pages tend toward the crassly commercial and are more extreme than one would likely find in this proportion of the top 100 (or even 200) of the results of a Google query for a medical condition.

2.1.2 MMED100 Corpus

The MMED1000 corpus was created on November 5th and 8th, 2004 by automatically downloading from Google the top 100 search results for each of the following 10 queries:

- Adrenoleukodystrophy
- Alzheimer's
- Endometriosis
- Fibromyalgia
- Obesity
- Pancreatic cancer
- Colloidal Silver
- Irritable Bowel Syndrome
- Late Lyme Disease
- Lower Back Pain

The queries were chosen to provide a broad range of what might be typical queries for health consumers on the web and the types of pages that would result from these queries.

Colloidal Silver was chosen in the hopes of providing a sufficient number of pages of questionable reliability. Adrenoleukodystrophy, Pancreatic Cancer, Alzheimer’s and Obesity were chosen because there is general agreement in the medical community that these are diseases or health issues, and on diagnostic techniques. They also cover a spectrum of occurrence rates, with Adrenoleukodystrophy being relatively rare and Obesity being relatively common. The other five queries were chosen because there is less agreement in both the medical community and the general population about the existence, frequency, severity and treatment of these conditions. In particular, Fibromyalgia and IBS can be exclusionary diagnoses without clear and successful treatment options, which can open the door to web pages with a range of questionable treatments.

For annotation purposes a subset of this corpus, MMED100, with 100 pages, was created by randomly selecting ten documents from each of the ten queries.

At this time neither corpus is publicly available. However they can be provided on request and it is anticipated that they will be made publicly available once a viable standard is established for the annotations.

2.2 Definitions

The primary classification task is to classify pages based on their reliability (quality or trustworthiness of the information they contain). The secondary classification task is to classify pages based on their type (e.g. commercial, patient leaflet, link). The classification by type emerged from the hypothesis that different types of pages may need to be treated differently to classify them based on their reliability. For example, if the primary purpose of a page is to provide links to information, determining the reliability of the page may require determining the reliability of the pages to which it links. However, in the current study, annotators are provided only the given web page and not allowed to follow links, so their reliability determination was made based on the apparent balance and objectivity of the links on the page.

For both tasks, only one tag was allowed, so annotators were instructed to consider the main purpose or intent of the page.

2.2.1 Reliability

Reliability of web pages is annotated based on a five level scale.

Probably Reliable (PrR)
The information on these pages appears to be complete and correct, meeting the standards of Evidence-Based Medicine where appropriate. Information is presented in a balanced and objective manner, with the full range of options discussed (where appropriate). The page and author appear reputable, with no obvious conflicts of interest. The appropriate disclaimers, policies, and contact information are present. Where appropriate, sources are cited. An example of a page in this category would be a patient leaflet from a reputable source that adheres to the standards of Evidence-Based Medicine.

Possibly Reliable (PoR)
The information on the page is generally good and without obvious false or outdated statements, but may not be sufficiently complete and balanced or may not conform to evidence-based standards. An example of a page in this category would be a patient leaflet that contains only a brief description of diagnostic procedures or suggests a treatment option that is generally accepted, but not supported by evidence.

Unable to determine (N)
For these pages it is difficult or impossible to determine the reliability, generally because there is not enough information. For example, the page may be blank, only contain login information, or be the front page of a medical journal.

Possibly Unreliable (PoU)
These pages may contain some reliable information, but either have some that is outdated, false or misleading, or the information is sufficiently unbalanced so as to be somewhat misleading. An example of a page that might fall into this category is a practitioner commercial pages, which has valid information about an illness, but only discuss the preferred treatment offered by the practitioner.

Probably Unreliable (PrU)
These pages contain false or misleading information, or present an unbalanced or biased viewpoint on the topic. Examples of pages in this category would include: testimonials (unsupported viewpoints or opinions of a single individual) or pages that are clearly promoting and selling a single treatment option.

2.2.2 Type of Page

We found six types of pages that frequently come up in search results for queries in the medical domain: Commercial, Patient Leaflet, Link, Medical Articles, Support, and Testimonials. There are also pages which are not relevant, or do not contain sufficient information to make a determination. Below we discuss each of these types. When a page seems to overlap categories the annotation is based on the primary purpose of the page.

Commercial (C)
The primary purpose of these pages is to sell something, for example, pages about an ailment sponsored by a drug (also more general treatment or equipment) company, which sells a drug to treat it. Given the desire to sell, these pages might not present complete or balanced information (making them less likely to be reliable). Practitioner pages with no real (substantial) information, which are designed to get people to make an appointment, as opposed to patient leaflets (designed to supplement information that patients receive in the office or clinic), might also fall into this category

Link (L)
The primary purpose of these pages is to provide links to other pages or sites (external), which will provide information about a certain illness or medical condition. These links may or may not be annotated, and the degree of annotation may vary considerably. Since the reliability of these pages depends on the reliability of the pages they link to (possibly also on the text in the annotations), without following the links a reliability estimate can be based on the range and apparent objectivity of the links.

Patient Leaflet, Brochure, Fact Sheet or FAQ (P)
The primary purpose of these pages is to provide information to patients about a specific illness or medical condition. Generally, these pages will be produced by a clinic, medical center, physician, or government agency, etc. The primary purpose is to provide information. This class needs to be distinguished from medical articles, especially in encyclopedias or the Merck Manual, etc. These pages will tend to have headings like: symptoms, diagnosis, treatment, etc. These headings can take the form of links to specific parts of the same page or to other pages on the same site (internal). The reliability of these pages is based on their content and determined by factors including Evidence-Based Medicine, completeness, and the presence of incorrect or outdated information.

Medical Article (practitioner or consumer) (MA)
The primary purpose of these pages is to discuss an aspect of a specific illness or medical condition, or a specific illness or medical condition. These can be divided into two main categories: articles aimed at consumers and articles aimed at health practitioners.

Articles aimed at health practitioners, particularly doctors, may be scientific research articles. The reliability of these pages is based on their content and determined by factors including Evidence Based Medicine, completeness, and the presence of incorrect or outdated information. Note: Medline search results may be considered a links page to medical articles.

Articles aimed at consumers may come from a variety of sources including mainstream and alternative media sources. Reliability is determined based on the content as with articles for practitioners.

Testimonial (T)
The primary purpose of these pages is to provide testimonial(s) of individuals about their experience with an illness, condition, or treatment. While individuals may be considered reliable when discussing their own personal experiences, these pages tend to be unreliable, because they are generally not objective or balanced. There is a tendency for readers to generalize from very specific information or experiences provided by the testimonial, which can be misleading.

Support (S)
The primary purpose of these pages is to provide support of sufferers (or their loved ones or caregivers) of a particular illness or condition. The pages may contain information, similar to that found in a patient leaflet; links to other sites, similar to a links page; and testimonials. In addition they may contain facilities such as chat rooms, newsletters, and email lists. Activities may include lobbying for funding for research, generally put up by individuals or non-profit organizations. For reliability, one may need to look at the agenda of the authors or group. It may be in their interest (politically) to overstate the problem or make things out to be worse then they are to secure increased funding or sympathy for their cause.

Not Relevant (N)
These pages are blank or not relevant and include: login pages, conditions of use pages, and medical journal front pages.

2.3 Annotation Study

In order to get started with system development, a single annotator, M, who was involved with development of both the classifications and the system, tagged the IBS70 and MMED100. Then in Spring 2008 two senior undergraduate science majors (chemistry and biology), L and E, were hired for the annotation study. The annotation study consisted to two primary phases: training and testing. Each phase is described below.

2.3.1 Training Phase

The two student annotators, L and E, received copies of the draft annotation instructions. They each met individually with M to discuss the instructions and any questions they had.

For each of three training runs, ten randomly chosen web pages from the IBS70 corpus were posted on a private web site. The students annotated the pages for reliability and type and then met individually to discuss their annotations with M. As questions and issues arose, the instructions were amended to reflect clarifications. For example, L needed additional instructions on the distinction between Link and Patient Leaflet pages; a separate category for FAQs was collapsed into the Patient Leaflet category.

2.3.2 Testing Phase

Once the student annotators seemed to be achieving reasonable levels of agreement (Cohen's Kappa above 0.4) on each task, there was a three-part testing phase. The remaining 40 pages in the IBS70 corpus were randomly divided into two test corpora and finally the MMED100 corpus was annotated.

During the testing phase, one of the students, L, seemed to annotate less carefully. (Possibly because the timing coincided with graduation and summer vacation.) For example, on the MMED100 corpus L tagged 30% as N (unable to determine the reliability, compared to 12% for E and 10% for M. L was asked to go back and reconsider the web pages tagged as N. We report results with L's reconsidered tags here for completeness, but further discussion will focus on agreement between M and E.

2.4 Testing and Analysis

We report inter-rater agreement using accuracy, Cohen's Kappa statistic (Cohen, 1960) for chance corrected agreement and F-Measure (Hripcsak and Rothschild, 2005). We consider each annotation axis separately.

2.4.1 Page Reliability

We can estimate a baseline distribution of the categories R (reliable), N (unable to determine), and U (unreliable) based on an average of the tags across all training and test sets to be approximately: 68% R; 13% N; 19% U.

Table 1 shows the results for the Accuracy (percent agreement) and Kappa statistic on the five reliability classes across all the corpora. It became immediately clear the annotators were not able to make the more fine-grained distinctions between "probably" and "possibly" for either the reliable or unreliable classes, given the current instructions and timeline. The classes were then collapsed to three: R (reliable), N (unable to determine) and U (unreliable) and the results are shown in Table 2.

Accuracy/ Kappa	5 Classes Reliability		
Set\Raters	M-E	M-L	E-L
IBS train	**0.47**/0.30	**0.33**/0.12	**0.40**/0.19
IBS test	**0.33**/0.11	**0.40**/0.25	**0.43**/0.28
MMed100	**0.51**/0.32	**0.35**/0.12	**0.38**/0.14

Table 1. Inter-rater agreement for 5-class reliability.

Accuracy/ Kappa	3 Classes Reliability		
Set\Raters	M-E	M-L	E-L
IBS train	**0.70**/0.44	**0.60**/0.25	**0.67**/0.33
IBS test	**0.70**/0.43	**0.65**/0.42	**0.75**/0.59
MMed100	**0.77**/0.49	**0.66**/0.30	**0.62**/0.22

Table 2. Inter-rater agreement for 3-classs reliability.

The results in Table 2 for M-E show improved agreement after training and consistent moderate agreement on the test corpora based on the Kappa statistic. Accuracy (percent agreement) for M-E is 70% for both IBS testing and training and 77% for the MMED100.

Further analysis of L's reliability tags showed a bias toward the "U" tag. For example, in the MMED100 corpus, L tagged 28% as U, compared to 19% and 17% for M and E, respectively.

Hripcsak and Rothschild (2005) suggest use of the F-measure (harmonic average of precision – equivalent to positive predictive value - and recall – equivalent to sensitivity - commonly used in Information Retrieval) to calculate inter-rater agreement in the absence of a gold standard. In Table 3 we report the average F-measure between each pair of raters and the F-measure by class. A higher F-measure indicates better agreement, so these results show that the "Can't Tell" class is the most difficult to agree on, followed by the "Unreliable" class.

MMED100 F-Measure	3 Classes Reliability		
Class\Raters	M-E	M-L	E-L
Reliable	0.87	0.78	0.76
Can't Tell	0.45	0.22	0.30
Unreliable	0.55	0.46	0.36
Average	**0.62**	**0.49**	**0.47**

Table 3. F-measure by class for 3-classs reliability.

In order to look for patterns of agreement between the raters we looked at agreement by query in the MMED100 corpus. In Table 4 we show the agreement for M and E by query. Although it appears that some queries were easier to annotate than others, since there are only 10 pages per query, the sample may be too small to draw definite conclusions.

Query	Accuracy	Kappa
Endometriosis	1	1
Pancreatic Cancer	1	1
Late Lyme	1	1
Adrenoleukodystrophy	0.8	0.412
Obesity	0.8	0.655
Alzheimer's	0.7	-0.154
Fibromyalgia	0.7	0.444
Lower Back Pain	0.7	-0.154
Colloidal Silver	0.6	0.13
Irritable Bowel Syndrome	0.4	-0.053

Table 4. Inter-rater reliability agreement for M-E by query.

Possible ways to improve these results are presented in the "Discussion" section.

2.4.2 Page Type

The dominant page types are P (patient leaflets), L (link), C (commercial) and MA (medical article). The baseline distribution based on averages across the training and test sets is approximately: 39% P; 15% L; 18% C; and 13% MA. The other three classes S (support), T (testimonial), and N (unable to determine) making up only 15% of the pages in the corpus.

Table 5 shows the results for Accuracy and the Kappa statistic on the seven type classes across all the corpora. Collapsing categories for the type annotation task did not appreciably increase Kappa scores (M-E Kappa was 0.742 on the MMED100 corpus when the P and MA classes were collapsed), so it seems preferable to keep the original classes.

Accuracy/ Kappa	Type		
Set\Raters	M-E	M-L	E-L
IBS train	**0.57**/0.42	**0.83**/0.78	**0.47**/0.28
IBS test	**0.73**/0.64	**0.65**/0.55	**0.73**/0.64
MMed100	**0.81**/0.73	**0.48**/0.29	**0.50**/0.31

Table 5. Inter-rater agreement for type annotation.

Again we see with annotators M and E, the improved agreement from training to testing, as distinctions between classes were clarified (for example, between Link and Patient Leaflets, and between Patient Leaflets and Medical Articles).

We also computed F-measure by type for the MMED100 corpus, as shown in Table 6. Of the three most common types of pages (Patient Leaflet, Link, Commercial), the Link type was the most difficult for M-E to agree on.

MMED100 F-Measure	Type		
Class\Raters	M-E	M-L	E-L
P	0.893	0.593	0.625
L	0.625	0.480	0.435
C	0.727	0.323	0.414
S	0.769	0.222	0.250
T	0.500	0.000	0.800
MA	0.667	0.593	0.455
N	0.857	0.143	0.118
Average	**0.720**	**0.336**	**0.442**

Table 6. F-measure by class for page type.

We further analyzed the page type annotations by query for raters M and E (Table 7). We found a negative correlation between the variance of the types in a query to the Kappa statistic of agreement for the query ($r^2 = -0.62$).

Query	Accuracy	Kappa
Endometriosis	0.9	0.851
Fibromyalgia	0.9	0.846
Alzheimer's	0.8	0.75
Irritable Bowel Syndrome	0.8	0.73
Obesity	0.8	0.697
Pancreatic Cancer	0.8	0.63
Colloidal Silver	0.8	0.63
Adrenoleukodystrophy	0.8	0.512
Lower Back Pain	0.8	0.512
Late Lyme	0.7	0.483

Table 7. Inter-rater type agreement for M-E by query.

3 Discussion

Librarians, scholars, and information scientists have done significant work on the quality (reliability) of print, and more recently, web information (for example, see Cook 2001, Alexander and Tate 1999). It is important to distinguish quality (reliability) from credibility (e.g. Danielson 2005), which is based on the users view of the information. Here we are interested in the quality of the information itself.

In a relatively early study, Impicciatore et al. (1997) sampled web documents relating to fever in children and found the quality of the information provided to be very low. In 2002, Eysenbach et al. conducted a review of studies assessing the quality of consumer health information on the web. Of the 79 studies meeting their inclusion criteria (essentially appropriate scope and quantitative analysis), they found that 70% of the studies concluded that reliability of medical information on the Web is a problem.

To address the question of how to determine the quality of medical information on the web, Fallis and Frické (2002) empirically tested several proposed indicators and found that the standard indicators of quality for print media could not be directly translated to consumer medical information on the Web. Price and Hersh (1999) developed a semi-automated system to filter out low quality consumer medical web pages based on approximately 30 criteria.

Annotation studies have been discussed and conducted in the computational linguistics community for a variety of annotation tasks, including subjectivity (e.g. Weibe et al. 1999) and opinion (e.g. Somasundaran et al. 2008). Artstein and Poesio (2008) surveyed inter-coder agreement in computational linguistics, including Cohen's Kappa.

To ensure a "gold standard" for training machine learning algorithms to do automatic classification a number of approaches could be pursued: the production of bias-corrected tags as described by Weibe et al. (1999); a new study with "expert" annotators – having a stronger medical background – and additional training; ask annotators to use existing web tools (e.g. American Accreditation HealthCare Commission) to assess the page quality; systematically assess whether the noise introduced by moderate agreement levels will create problems for machine learning with this data (Beigman Klebanov and Beigman 2009).

The agreement on the type annotation task could still be improved, possibly by additional clarification to the definitions. However, it is still to be determined if noise levels are low enough and sufficiently random to be used successfully in supervised learning. This task is easier than the reliability task and requires less expertise of the annotators.

4 Conclusion

There is a demonstrated need to provide tools to health care consumers to automatically filter web pages by the reliability, quality, or trustworthiness of the medical information the pages contain. We have shown promising results in this study that appropriate classes of pages can be developed. These classes can be used by human annotators to annotate web pages with reasonable to good agreement.

Thus we have laid a foundation for future annotation studies to create a gold standard data set of consumer medical web pages. The corpora in this study are currently being used to create an automated system to estimate the reliability of medical web pages.

Acknowledgments

This work was supported in part by a CSU Stanislaus Naraghi Faculty Research Enhancement Grant. I am grateful to Elizabeth Jimenez and Luis Adalco for participating in the annotation study and to the anonymous reviews for their comments and suggestions. I would also like to thank Roger Hartley my dissertation advisor and Peter Foltz for discussions during the formulation and development of the system, and Tom Carter for helpful and insightful comments leading to the improvement of this paper.

References

Janet E. Alexander and Marsha Ann Tate. 1999. *Web Wisdom: How to Evaluate and Create Information Quality on the Web.* Lawrence Erlbaum and Associates, New Jersey.

American Accreditation HealthCare Commission. *Health information on the internet: A checklist to help you judge which websites to trust.* Retrieved February 28, 2010, from http://www.urac.org

R. Artstein and M. Poesio. 2008. Inter-coder agreement for computational linguistics. *Comput. Linguist.* 34, 4 (Dec. 2008), 555-596.

B. Beigman Klebanov, and E. Beigman. 2009. From annotator agreement to noise models. *Comput. Linguist.* 35, 4 (Dec. 2009), 495-503.

J. Cohen. 1960. A coefficient of agreement for nominal scales. *Educational and Psychological Measurement*, 20, pages 34-46.

Alison Cooke. 2001. *A Guide to Finding Quality Information on the Internet: Selection and Evaluation Strategies, Second Edition.* Library Association Publishing, London.

D.R. Danielson. 2005. Web credibility. C. Ghaoui (Ed.), *Encyclopedia of Human-Computer Interaction.* Hershey, PA: Idea Group, 713-721.

Gunther Eysenbach, John Powell, Oliver Kuss, and Eun-Ryoung Sa. 2002. Empirical Studies Assessing the Quality of Health Information for Consumers on the World Wide Web: A Systematic Review. *JAMA*, May 22, 2002; 287(20): 2691 - 2700.

Don Fallis and Martin Frické. 2002. Indicators of Accuracy of Consumer Health Information on the Internet. *Journal of the American Medical Informatics Association,* 9, 1, (2002): 73-79.

George Hripcsak and Adam S. Rothschild. 2005. Agreement, the F-Measure, and Reliability in Information Retrieval. *J Am Med Inform Assoc.* 2005 May–Jun; 12(3): 296–298.

Piero Impicciatore, Chiara Pandolfini, Nicola Casella, and Maurizio Bonati. 1997. Reliability of Health Information for the Public on the World Wide Web: Systematic Survey of Advice on Managing Fever in Children at Home. *BMJ* 1997; 314:1875 (28 June).

Melanie J. Martin. 2004. Reliability and Verification of Natural Language Text on the World Wide Web. Paper at *ACM-SIGIR Doctoral Consortium*, July 25, 2004, Sheffield, England.

Susan L. Price and William R. Hersh. 1999. Filtering Web Pages for Quality Indicators: An Empirical Approach to Finding High Quality Consumer Health Information. *American Medical Informatics Association* 1999.

Swapna Somasundaran, Josef Ruppenhofer and Janyce Wiebe. 2008. Discourse Level Opinion Relations: An Annotation Study. Proceedings of the 9th SIGdial Workshop on Discourse and Dialogue Columbus, Ohio, June 19-20, 2008, pp. 129–137.

Janyce M. Wiebe, Rebecca F. Bruce, and Thomas P. O'Hara. 1999. Development and use of a gold-standard data set for subjectivity classifications. In *Proceedings of the 37th Annual Meeting of the Association For Computational Linguistics on Computational Linguistics* (College Park, Maryland, June 20 - 26, 1999). Annual Meeting of the ACL. Association for Computational Linguistics, Morristown, NJ, 246-253.

Automated Identification of Synonyms in Biomedical Acronym Sense Inventories

Genevieve B. Melton
Institute for Health Informatics & Dept of Surgery
University of Minnesota
Minneapolis, MN 55455 USA
gmelton@umn.edu

SungRim Moon
Institute for Health Informatics
University of Minnesota
Minneapolis, MN 55455 USA
moonx086@umn.edu

Bridget McInnes
College of Pharmacy
University of Minnesota
Minneapolis, MN 55455 USA
bthomson@umn.edu

Serguei Pakhomov
College of Pharmacy
University of Minnesota
Minneapolis, MN 55455 USA
pakh0002@umn.edu

Abstract

Acronyms are increasingly prevalent in biomedical text, and the task of acronym disambiguation is fundamentally important for biomedical natural language processing systems. Several groups have generated sense inventories of acronym long form expansions from the biomedical literature. Long form sense inventories, however, may contain conceptually redundant expansions that negatively affect their quality. Our approach to improving sense inventories consists of mapping long form expansions to concepts in the Unified Medical Language System (UMLS) with subsequent application of a semantic similarity algorithm based upon conceptual overlap. We evaluated this approach on a reference standard developed for ten acronyms. A total of 119 of 155 (78%) long forms mapped to concepts in the UMLS. Our approach identified synonymous long forms with a sensitivity of 70.2% and a positive predictive value of 96.3%. Although further refinements are needed, this study demonstrates the potential value of using automated techniques to merge synonymous biomedical acronym long forms to improve the quality of biomedical acronym sense inventories.

1 Introduction

Acronyms and abbreviations are increasingly used in biomedical text. This is in large part due to the expansive growth of the biomedical literature estimated to be close to one million articles annually (Stead et al. 2005). Ambiguous acronyms represent a challenge to both human readers and computerized processing systems for resolving the acronym's meaning within a particular context. For any given acronym, there are often multiple possible long form expansions. Techniques to determine the context-specific meaning or sense of an ambiguous acronym are fundamentally important for biomedical natural language processing and can assist with important tasks such as information retrieval and information extraction (Friedman 2000).

Acronym ambiguity resolution represents a special case of word sense disambiguation (WSD) with unique challenges. In particular, there are increasing numbers of new acronyms (i.e., short forms) as well as increasing numbers of new senses (i.e., long forms) for existing acronyms within biomedical text. Acronyms in biomedicine also range from those that are common, to those that are infrequent which appear to be created in an ad hoc fashion resulting essentially in neologisms distinct to small sets of biomedical discourse.

Sense inventories are important tools that can assist in the task of disambiguation of acronyms and abbreviations. The relative formal nature of biomedical literature discourse lends itself well to building these inventories because long forms are typically contained within the text itself, providing a "definition" on its first mention in an article, next to a parenthetical expression containing the short form or vice versa (Schwartz and Hearst 2003). In contrast, clinical documents are less structured and

Proceedings of the NAACL HLT 2010 Second Louhi Workshop on Text and Data Mining of Health Documents, pages 46–52,
Los Angeles, California, June 2010.

typically lack expanded long forms for acronyms and abbreviations, leaving sense inventories based on documents in the clinical domain not as well developed as the sense inventories developed from the biomedical literature (Pakhomov et al. 2005).

Compilation of sense inventories for acronyms in clinical documents typically relies on vocabularies contained in the Unified Medical Language System (UMLS) as well as other resources such as ADAM (Zhou et al. 2006). However, with the advantage of using rich and diverse resources like ADAM and the UMLS comes the challenge of having to identify and merge synonymous long form expansions which can occur for a given short form. Having synonymous long forms in a sense inventory for a given acronym poses a problem for automated acronym disambiguation because the sense inventory dictates that the disambiguation algorithm must be able to distinguish between semantically equivalent senses. This is an important problem to address because effective identification of synonymous long forms allows for a clean sense inventory, and it creates the ability for long form expansions to be combined while preserving the variety of expression occurring in natural language. By automating the merging of synonymous expansions and building a high quality sense inventory, the task of acronym disambiguation will be improved resulting in better biomedical NLP system performance.

Our approach to reducing multiple synonymous variants of the same long form for a set of ten biomedical acronyms is based on mapping sense inventories for biomedical acronyms to the UMLS and using a semantic similarity algorithm based on conceptual overlap. This study is an exploratory evaluation of this approach on a manually created reference standard.

2 Background

2.1 Similarity measures in biomedicine

The area of semantic similarity in biomedicine is a major area within biomedical NLP and knowledge representation research. Semantic similarity aids NLP systems, improves the performance of information retrieval tasks, and helps to reveal important latent relationships between biomedical concepts. Several investigators have studied conceptual similarity and have used relationships in controlled biomedical terminologies, empiric statistical data from biomedical text, and other knowledge sources (Lee et al. 2008; Caviedes and Cimino 2004). However, most of these techniques focus on generating measures between a single pair of concepts and do not deal directly with the task of comparing two groups of concepts.

Patient similarity represents an important analogous problem that deals with sets of concepts. The approach used by Melton et al. (2006) was to represent each patient case as a set of nodes within a controlled biomedical terminology (SNOMED CT). The investigators then applied several measures to ascertain similarity between patient cases. These measures ranged from techniques independent of the controlled terminology (i.e. set overlap or Hamming distance) to methods heavily reliant upon the controlled terminology based upon path traversal between pair of nodes using defined relationships (either IS-A relationships or other semantic relationships) within the terminology.

2.2 Lesk algorithm for measuring similarity using sets of definitional words

A variety of techniques have been used for the general problem of WSD that range from highly labor intensive that depend upon human data tagging (i.e. supervised learning) to unsupervised approaches that are completely automated and rely upon non-human sources of information, such as context and other semantic features of the surrounding text or definitional data.

The Lesk algorithm (Lesk 1986) is one example of an unsupervised method that uses dictionary information to perform WSD. This algorithm uses the observation that words co-occurring in a sentence refer to the same topic and that dictionary definition words will have topically related senses, as well. The classic form of this algorithm returns a measure of word overlap. Lesk depends upon finding common words between dictionary definitions. One shortcoming of Lesk, however, it that it can perform worse for words with terse, few word definitions.

As a modification of Lesk, researchers have proposed using WordNet (Felbaum 1998) to enhance its performance. WordNet has additional semantic information that can aid in the task of disambiguation, such as relationships between the term of interest and other terms. Banerjee and Pedersen (2002) demonstrated that modifications to Lesk improved performance significantly with the addition of semantic relationship information.

2.3 Biomedical literature sense inventories

A number of acronym and abbreviation sense inventories have been developed from the biomedical literature using a variety of approaches. Chang et al. (2002) developed the Stanford biomedical abbreviation server[1] using titles and abstracts from MEDLINE, lexical heuristic rules, and supervised logistic regression to align text and extract short form/long form pairs that matched well with acronym short form letters. Similarly, Adar (2004) developed the Simple and Robust Abbreviation Dictionary (SaRAD)[2]. This inventory, in addition to providing the abbreviation and definition, also clusters long forms using an N-gram approach along with classification rules to disambiguate definitions. This resource, while analogous with respect to its goal of merging and aligning long form expansions, is not freely available. Adar measured a normalized similarity between N-gram sets and then clustered long forms to create a clustered sense inventory resource.

One of the most comprehensive biomedical acronym and abbreviation databases is ADAM (Zhou et al. 2006) an open source database[3] that we used for this study. Once identified, short form/long form pairs were filtered statistically with a rule of length ratio and an empirically-based cut-off value. This sense inventory is based on MEDLINE titles and abstracts from 2006 and consists of over 59 thousand abbreviation/long form pairs. The authors report high precision with ADAM (97%) and up to 33% novel abbreviations not contained within the UMLS or Stanford Abbreviation dictionary.

2.4 MetaMap resource for automated mapping to the UMLS

An important resource for mapping words and phrases to the UMLS Metathesaurus is MetaMap. This resource was developed at the National Library of Medicine (Aronson 2001) to map text of biomedical abstracts to the UMLS. MetaMap uses a knowledge intensive approach that relies upon computational linguistic, statistical, and symbolic/lexical techniques. While MetaMap was initially developed to help with indexing of biomedical literature, it has been applied and expanded successfully to a number of diverse applications including clinical text.

With each mapping, an evaluation function based upon centrality, variation, coverage, and cohesiveness generates a score for a given mapping from 0 to 1000 (strongest match). A cut-off score of 900 or greater is considered to represent a good conceptual match for MetaMap and was used in this study as the threshold to select valid mappings.

3 Methods

Ten randomly selected acronyms with between 10 to 20 long forms were selected from the ADAM resource database for this pilot study.

3.1 Long form mappings to UMLS

Each acronym long-form was mapped to the UMLS with MetaMap using two settings. First, MetaMap was run with its default setting on each long form expansion. Second, MetaMap was run in its "browse mode" (options "-zogm") which allows for term processing, overmatches, concept gaps, and ignores word order.

Processing each long form with MetaMap then resulted in a set of Concept Unique Identifiers (CUIs) representing the long form. Each CUI with a score over 900 was included in the overall set of CUIs for a particular long form expansion. For a given pair of long form expansions the two sets of CUIs that each long form mapped to were compared for concept overlap, in an analogous fashion to the Lesk algorithm. The overlap between concept sets was calculated between each pair of long form expansions and expressed as a ratio:

$$\frac{\#\ overlapping\ concepts\ shared\ between\ long\ forms}{\#\ concepts\ for\ the\ long\ form\ with\ least\ \#\ concepts}.$$

For this study, an overlap of 50% or greater was considered to indicate a potential synonymous pair.

Now let us assume that we have two concept sets: The first one is {A, B} and the second one is {A, B, C}, with each CUI having a score over 900. In this example, the overlap of concepts for the first concept set between it and the other is 100%, and for the second that is 66.7%. Because overlaps

[1] http://abbreviation.stanford.edu
[2] http://www.hpl.hp.com/shl/projects/abbrev.html
[3] http://arrowsmith.psych.uic.edu

are greater than 50%, they are a potential synonymous pair, and the overlap ratio is calculated as $\frac{\#\{A,B\}}{\#\{A,B\}} = \frac{2}{2} = 1$ (100%).

3.2 Expert-derived reference standard

Two physicians were asked to judge the similarity between each pair combination of long forms expansions on a continuous scale for our initial reference standard. Physicians were instructed to rate pairs of long forms for conceptual similarity. Long forms were presented on a large LCD touch-screen display (Hewlett-Packard TouchSmart 22" desktop) along with a continuous scale for the physicians to rate long form pairs as dissimilar (far left screen) or highly similar (far right screen). The rating was measured on a scale from 1 to 1500 pixels representing the maximum width of the touch sensitive area of the display (along the x-coordinate). Inter-rater agreement was assessed using Pearson correlation.

Expert scores were then averaged and plotted on a histogram to visualize expert ratings. We subsequently used a univariate clustering approach based on the R implementation of the Partitioning Around Medoids (PAM) method to estimate a cut-off point between similar and dissimilar terms based on the vector of the average responses by the two physicians. The responses were clustered into two and three clusters based on an informal observation of the distribution of responses on the histogram showing evidence of at least a bimodal and possibly a trimodal distribution.

As a quality measure, a third physician manually reviewed the mean similarity ratings of the first two physicians to assess whether their similarity judgments represented the degree of synonymy between long form expansions necessary to warrant merging the long form expansions. This review was done using a binary scale (0=not synonymous, 1=synonymous).

3.3 Evaluation of automated methods

Long form pair determinations based on the mappings to the UMLS were compared to our reference standard as described in Section 3.2. We calculated overall results of all long form pair comparisons and on all long form pairs that mapped to the UMLS with MetaMap. Performance is reported as sensitivity, specificity, and positive predictive value.

4 Results

A total of 10 random acronyms were used in this study. All long forms for these 10 acronyms were from the sense inventory ADAM (Zhou et al., 2006). This resulted in a total of 155 long form expansions (median 16.5 per acronym, range 11-19) (Table 1).

Acronym	N of LF expansions	LF expansions mapped by MetaMap
Total	155	119 (78%)
ALT	13	9 (70%)
CK	14	9 (64%)
CSF	11	7 (74%)
CTA	19	14 (74%)
MN	19	17 (89%)
NG	17	15 (88%)
PCR	17	8 (47%)
PET	17	15 (88%)
RV	16	14 (88%)
TTP	12	11(92%)

Table 1. Number of acronym long forms in ADAM and mapping to the UMLS

4.1 Long form mappings to UMLS

The default mode of MetaMap resulted in 119 (78%) long forms with mappings to the UMLS with MetaMap (Table 1). Use of MetaMap's browse mode did not increase the total number of mapped long forms but did change some of the mapped concepts returned by MetaMap (not depicted).

Acronym	N pairs	Pearson r
Total	1125	0.78*
ALT	78	0.79*
CK	91	0.77*
CSF	55	0.80*
CTA	136	0.92*
MN	171	0.69*
NG	136	0.68*
PCR	136	0.89*
PET	136	0.78*
RV	120	0.67*
TTP	66	0.76*

Table 2. Pearson correlation coefficient for ratings overall and for individual acronyms. *p<0.0001

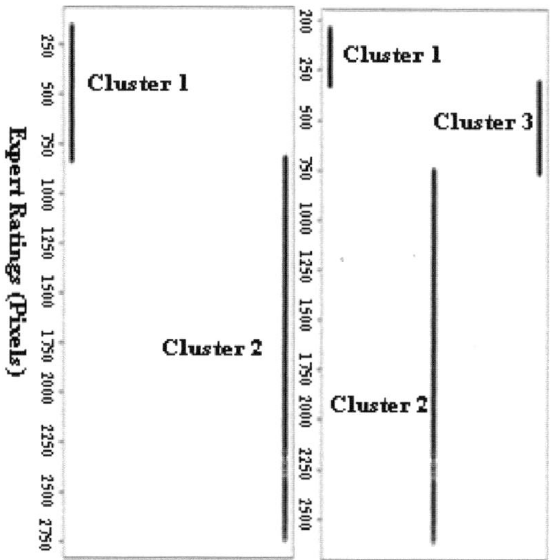

Figure 1. Two-way and three-way clustering solution of expert ratings of long form pairs.

4.2 Expert-derived reference standard

For the 1125 total comparison pairs, two raters assessed similarity between long form pairs on a continuous scale. The overall mean correlation between the two raters was 0.78 (standard deviation 0.08). Pearson correlation coefficients for each acronym are depicted in Table 2.

Two-way and three-way clustering demonstrated an empirically determined "cutoff" of 525 pixels from the left of the screen. This separation point between clusters (designated as "low cutoff") was evident on both the two-way and three-way clustering approaches using the PAM method to estimate a cut-off point between similar and dissimilar terms based on the vector of the average responses by the two physicians (Figure 1). Intuitively this low cutoff includes manual ratings indicative of moderate to low similarity (as 525 pixels along a 1500 pixel-wide scale is approximately one-third of the way from the left "dissimilar" edge of the touch-sensitive screen). To isolate terms that were rated as highly similar, we also created an arbitrary "high cutoff" of 1200 pixels.

CTA:	
"CT hepatic arteriography"	"CT angiography"
MN:	
"median nerve"	"motor neuron"
RV:	
"rabies virus"	"rotavirus"
"right ventricular free wall"	"right ventricle"
TTP:	
"thiamine triphosphate"	"thymidine triphosphate"

Figure 2. Examples of terms originally rated as highly similar but not synonymous by the curating physician.

Expert curation of the ratings by the third physician demonstrated that conceptual similarity ratings were sometimes not equivalent to synonymy that would warrant the collapse of long form pairs. Of 1125 total pairs of long forms, 70 (6%) origi-

	Default Mode: MetaMap		Browse Mode: MetaMap	
	All LF	Mapped LF only	All LF	Mapped LF only
High Cutoff				
Sensitivity	21.6%	39.6%	23.8%	43.8%
Specificity	98.1%	96.8%	99.4%	99.0%
PPV	48.7%	48.7%	77.8%	77.8%
NPV	93.6%	95.5%	93.9%	95.9%
Expert Curation				
Sensitivity	34.3%	64.9%	37.1%	70.2%
Specificity	98.6%	97.7%	99.9%	99.8%
PPV	61.5%	61.5%	96.3%	96.3%
NPV	95.8%	98.0%	96.0%	98.3%

Table 3. Performance of automated techniques for merging biomedical long form senses for all long forms and for long forms that mapped to the UMLS only.
PPV, positive predictive value; NPV, negative predictive value.

nally classified as similar were re-classified as conceptually different by the third physician. Several examples of long form pairs that were originally rated as highly similar but were judged as not synonymous are contained in Figure 2.

4.3 Evaluation of automated methods

The performance of our algorithm is shown in Table 3 using MetaMap in the default mode and browse mode and then applying our reference standard using the "low cutoff", "high cutoff", and expert curation (Table 3). Performance is reported for all 155 long forms (All LF) and for the subset of 119 long forms that mapped to the UMLS (Mapped LF only). Compared to the "low cutoff" reference standard, the "high cutoff" and expert curation were positively associated with more consistent performance. The browse mode identified fewer potential terms to merge and had higher accuracy than the default MetaMap mode.

5 Conclusions

The results of this pilot study are promising and demonstrate high positive predictive value and moderate sensitivity for our algorithm, which indicates to us that this technique with some additional modifications has value. We found that mapping long form expansions to a controlled terminology to not be straightforward. Although approximately 80% of long forms mapped, another 20% were not converted to UMLS concepts. Because each long form resulted in multiple paired comparisons, a 20% loss of mappings resulted globally in a 40% loss in overall system performance. While long form expansions were entered into MetaMap using a partially normalized representation of the long form, it is possible that additional normalization will improve our mapping.

An important observation from our expert-derived reference standard was that terms judged by physicians as semantically highly similar may not necessarily be synonymous (Figure 2). While semantic similarity is analogous, there may be some fundamentally different cognitive determinations between similarity and synonymy for human raters.

The current technique that we present compares sets of mapped concepts in an analogous fashion to the Lesk algorithm and other measures of similarity between groups of concepts previously reported. This study did not utilize features of the controlled terminology nor statistical information about the text to help improve performance. Despite the lack of additional refinement to the presented techniques, we found a flat overlap measure to be moderately effective in our evaluation.

6 Future Work

There are several lines of investigation that we will pursue as an extension of this study. The most obvious would be to use semantic similarity measures between pairs of concepts that capitalize upon features and relationships in the controlled terminology. We can also expand upon the type of similarity measures for the overall long form comparison which requires a measure of similarity between *groups* of concepts. In addition, an empiric weighting scheme based on statistical information of common senses may be helpful for concept mappings to place more or less emphasis on important or less important concepts. We plan to determine the impact of automatically reduced sense inventories on the evaluation of WSD algorithms used for medical acronym disambiguation.

Finally, we would like to utilize this work to help improve the contents of a sense inventory that we are currently developing for acronyms and abbreviations. This sense inventory is primarily based on clinical documents but incorporates information from a number of diverse sources including ADAM, the UMLS, and a standard medical dictionary with abbreviations and acronyms.

Acknowledgments

This work was supported by the University of Minnesota Institute for Health Informatics and Department of Surgery and by the National Library of Medicine (#R01 LM009623-01). We would like to thank Fairview Health Services for ongoing support of this research.

References

Eytan Adar (2004) SaRAD: A simple and robust abbreviation dictionary. Bioinformatics 20:527–33.

Alan R Aronson (2001) Effective mapping of biomedical text to the UMLS Metathesaurus: the MetaMap program. Proc AMIA Symp. 2001:17-21.

Satanjeev Banerjee, Ted Pedersen. 2002. An Adapted Lesk Algorithm for Word Sense Disambiguation Using WordNet, Proceedings of the Third International Conference on Computational Linguistics and Intelligent Text Processing, p.136-145, February 17-23.

Jorge E. Caviedes JE, James J Cimino. (2004) Towards the development of a conceptual distance metric for the UMLS. J Biomed Inform. Apr;37(2):77–85.

Jeffrey T Chang, Hinrich Schutze, Russ B Altman (2001) Creating an online dictionary of abbreviations from Medline. J Am Med Inform Assoc 9:612–20.

Christiane Fellbaum, editor. 1998. *WordNet: An electronic lexical database*. MIT Press.

Carol Friedman. 2000. A broad-coverage natural language processing system. Proc AMIA Symp., 270–274.

Wei-Nchih Lee, Nigam Shah, Karanjot Sundlass, Mark Musen (2008) Comparison of Ontology-based Semantic-Similarity Measures. AMIA Annu Symp Proc. 2008. 384–388.

Michael E. Lesk. 1986. Automatic sense disambiguation using machine readable dictionaries: How to tell a pine cone from a ice cream cone. In *Proceedings of SIGDOC '86*.

Genevieve B. Melton, Simon Parsons, Frances P. Morrison, Adam S. Rothschild, Marianthi Markatou, George Hripcsak. 2006. Inter-patient distance metrics using SNOMED CT defining relationships, Journal of Biomedical Informatics, 39(6), 697-705.

Serguei Pakhomov, Ted Pedersen, Christopher G. Chute. 2005. Abbreviation and Acronym Disambiguation in Clinical Discourse. American Medical Informatics Association Annual Symposium, 589-593.

Ariel S Schwartz and Marti A. Hearst. 2003. A Simple Algorithm for Identifying Abbreviation Definitions in Biomedical Text. Pacific Symposium on Biocomputing p451-462.

William W Stead, Brian J Kelly, Robert M Kolodner. 2005. Achievable steps toward building a National Health Information infrastructure in the United States. J. Am. Med. Inform. Assoc., 12, 113–120.

Wei Zhou, Vetle I Torvik, Neil R Smalheiser (2006) ADAM: Another database of abbreviations in Medline. Bioinformatics 22:2813– 8.

Characteristics and Analysis of Finnish and Swedish Clinical Intensive Care Nursing Narratives

Helen Allvin[f], Elin Carlsson[f], Hercules Dalianis[f], Riitta Danielsson-Ojala[a], Vidas Daudaravičius[b], Martin Hassel[f], Dimitrios Kokkinakis[c], Heljä Lundgren-Laine[a], Gunnar Nilsson[f], Øystein Nytrø[d], Sanna Salanterä[a], Maria Skeppstedt[f], Hanna Suominen[e], Sumithra Velupillai[f]

[a]Department of Nursing Science, University of Turku, VSSHP, Turku, Finland,
[b]Vytautas Magnus University, Lithuania,
[c]Department of Swedish, University of Gothenburg, Sweden,
[d]IDI, The Norwegian University of Science and Technology, Norway,
[e]NICTA Canberra Research Laboratory and Australian National University, Australia
[f]Department of Computer and Systems Sciences/Stockholm University
Forum 100
SE-164 40 Kista, Sweden
`http://www.dsv.su.se/hexanord`

Abstract

We present a comparative study of Finnish and Swedish free-text nursing narratives from intensive care. Although the two languages are linguistically very dissimilar, our hypothesis is that there are similarities that are important and interesting from a language technology point of view. This may have implications when building tools to support producing and using health care documentation. We perform a comparative qualitative analysis based on structure and content, as well as a comparative quantitative analysis on Finnish and Swedish Intensive Care Unit (ICU) nursing narratives. Our findings are that ICU nursing narratives in Finland and Sweden have many properties in common, but that many of these are challenging when it comes to developing language technology tools.

1 Introduction

The purpose of this study[1] is to do content and lexical analysis of nursing narratives written in an Intensive Care Unit (ICU). The ultimate goal of our research is to define linguistic similarities and language-specific aspects that differentiate clinical narratives in Finnish and Swedish in order to lay groundwork for developing internationally applicable language technology solutions and create a framework for characterising and comparing clinical narratives. Free text is handy for information entry but a challenge for information extraction, care handover and other uses of gathered information. Language technology can alleviate some of these problems in retrospective analysis by offering a more semantically informed interpretation and abstraction. However, the most promising potential of language technology is to interactively improve, interpret and code during text entry so that the resulting structured, coded, free text can be validated immediately. The critical bottleneck today is namely information handover and reuse, and extensive text is simply not used nor is useful. Interactively validated, semantically processed text could be more usable and support abstraction, visualization and query tools for the benefit of clinicians, patients, researchers and quality administrators.

[1] Our research on the Stockholm EPR Corpus (Dalianis et al., 2009) has been approved by Etikprövningsnämnden i Stockholm, (the Regional Vetting Board), reference number 2009/1742-31/5. Our research on the Finnish Corpus (Salanterä et al., 2009) has been approved by the Ethical committee of the Hospital District of South West Finland, reference number 2/2009, §66.

Proceedings of the NAACL HLT 2010 Second Louhi Workshop on Text and Data Mining of Health Documents, pages 53–60,
Los Angeles, California, June 2010.

In this paper, we analyze Finnish and Swedish ICU nursing narratives from both qualitative and quantitative perspectives. Our data includes textual nursing documentation of adult patients with a protracted inpatient period. We have chosen ICUs because of their international similarity in decision making (Lauri & Salanterä 2002) and nursing documentation because it covers the entire inpatient period.

2 Background

2.1 Clinical text

Clinical text covers the text documents produced for clinical work by clinicians and occurs in clinical information systems. It is written by *clinicians*, that is, professionals (physicians, nurses, therapists and other specialist) responsible for patient care. Its primary purpose is to serve patient care as a summary or hand-over note. However, clinical text is also written for legal requirements, care continuity and purposes of reimbursement, management and research. Clinical text covers every care phase and, depending on the purpose, documents differ. Documents that describe the patient's state, current health problems and socio-medical history are very different from those describing a care plan, its actualization and evaluation of care outcomes. Again, these differ from diagnostic notes, lab results, radiography readings, pathology reports and discharge documents that plan further care at discharge. Finally, clinical text may have been entered in "real time", in retrospect, or as a summary, by the bedside or elsewhere. The enterer can be a clinician, secretary who transcribes a dictate, speech recognition software or another system that generates or synthesizes text, (McDonald 1997, Thoroddsen et al., 2009.).

2.2 Legal requirements for clinical documentation in different countries

In several countries clinical documentation is based on law. In Finland, the Ministry of Social Affairs and Health (Statutes of Finland, 298/2009) defines that to ensure good care, all necessary and wide-ranging information has to be registered in patient records. In Sweden, the National Board of Health and Welfare has a similar approach (Patientdatalagen 2008:355). Clinical text should be explicit and intelligible, and only generally well-known, accepted concepts and abbreviations are allowed to be used. It should detail adequately the patient's conditions, care and recovery.

2.3 Special features of ICU and nursing

An ICU is an essential component of most large hospitals with high quality care. ICUs provide care for critically ill patients and focus on conditions that are life-threatening and require comprehensive care and constant monitoring (Webster's 2010). This task is fairly similar universally. It is based on optional, international guidelines focusing on triage, admission, discharge and education. This international similarity was evident when nurses' decision making was studied in Canada, Finland, Northern Ireland, Switzerland, Norway and the USA (Lauri & Salanterä 2002); the study showed that decision making of ICU nurses was the most uniform in different countries when compared with nurses working in public health care, psychiatric care, and short and long term care.

Clinical text written by nurses, that is, *nursing narratives*, both in Finland and in Sweden is based on the care process which stands for gathering information from the patient, setting goals for care, implementing nursing interventions, and evaluating the results of given care. In Finland, the national standardized documentation model has been implemented with the Finnish care classification (assessment, interventions and outcomes of care) (Tanttu & Ikonen 2007). The Swedish VIPS model provides a structure for the documentation process with key words that reflect the nursing process (Ehrenberg et al., 1996).

ICU nursing narratives can be lengthy, especially when the patient stay in the ICU is prolonged. As much as 60 A4 pages equivalents of written text may be gathered during one period of care. However, clinicians have somewhat different opinion on how to organize the information they write. For example, headings are often inconsistent and text under headings can cover a lot of other issues than those directly concerning the given heading. (Suominen et al., 2009.)

2.4 Related studies

Since most of the available clinical documents are in free-text form, a number of stylistically oriented efforts to characterize the data from various angles have taken place. This may include various topics, from viewing detailed information about specific items (e.g. readability, Kim et al., 2007) to identifying patterns and structures in order to provide better technology to automatically process the sublanguage (Pakhomov et al., 2006). The majority of such efforts investigate different aspects of linguistic features at a monolingual level, for instance, Hahn & Wermter (2004); Tomanek et al., (2007); Chung (2009); Harkema et al., (2009); while for a thorough review of various related issues see Meystre et al., (2008). In the Nordic context, Josefsson (1999) discusses Swedish clinical language and shows examples on how verb constructions in a clinical setting differ from a non clinical setting. One claim is that the physician unmarks the verb forms for agentivity when writing about the patient and what actions she takes, for example, *Patienten hallucinerar* [*The patient hallucinates*] instead of the normal form *Patienten får hallucinationer* [*The patient experiences hallucinations*].

Hellesø (2005) describes nurses' general use of the language function in the nursing discharge notes. She finds that the text in the nursing discharge notes is information-dense and characterized by technical terms, and that the use of standardised templates helped nurses improve the completeness, structure and content of the information. Comparisons at a monolingual level between written clinical text and lay text has been carried out by Dalianis et al., (2009). A contrastive computational linguistics study was carried out between the Stockholm EPR Corpus (SEPR) and a general language corpus, both written Swedish text. The findings showed that SEPR contained longer words and that the vocabulary was highly domain-specific. Other work is described in Ownby, (2005). Comparing clinical text at a crosslingual level has, to our knowledge, only been done by Borin et al., (2007).

3 Analysis of Finnish and Swedish ICU nursing narratives

The analyzed nursing narratives origin from one ICU in a university-affiliated hospital both in Finland and Sweden. Our inclusion criterion was an ICU inpatient period of at least 5 days and patient's age of at least 16 years. The Finnish data includes nursing narratives from 514 patient records (496 unique patients, 18 rebounds, a patient record is defined as each inpatient period of at least 5 days per patient) between January 2005 and August 2006. The Swedish data includes nursing narratives from 379 patient records (333 unique patients, 46 rebounds) between January 2006 and May 2008. Since we did not have complete admission and discharge documents from both countries, our analysis is performed on daily nursing narratives. These documents are written by ICU nurses during the actual inpatient period from the patient admission to the discharge.

3.1 Qualitative analysis

A manual content analysis was performed by four health care professionals (i.e., three native Finnish speakers knowing Swedish, one Swedish native speaker) and one native Swedish speaking language consultant. Three average-sized patient records each from Finland and Sweden were chosen for our analysis (average size 2,389 words for Finland and 5,169 words for Sweden). In the analysis, we considered special features (Table 1) of daily notes both from the structural and content related points of view.

The style and context of both Finnish and Swedish text is very similar. For health care professionals, and especially with an ICU background, all the texts are intelligible and the meaning of a writer becomes evident from the context even in the presence of numerous linguistic and grammatical mistakes; almost all the sentences are lacking both grammatical subjects and objects. It is evident that in both countries, the narratives are written from a professional to a professional in order to support information transfer, remind about important facts, and supplement numerical data.

A feature common for all the six records is that they rarely contain any subjects or objects when nurses are writing about patients. However, in the Swedish nursing narratives the word *patient* is used as a subject or object much more often than in the Finnish narratives. The abbreviation *pat.* is mostly used for this reference and *she/he* is never used for this purpose. In the whole data, *pat.* is 40 percent more common than *she/he*, which is the most common personal pronoun. It seems that the word *patient* or *pat.* is used more when the professionals are writing about relatives. In general, pronouns are used infrequently in the narratives, and

Special features of Finnish narratives		**Special features of Swedish narratives**	
Structure	**Examples**	**Structure**	**Examples**
Headings are used in 2 out of 3 patient records. Headings are typically used as subjects or subjects are partially used.	*Diuresis: occasionally profuse.* *(Diureesi: ajoittain runsasta.)* *Pupils move under eyelids but does not open eyes.* *(Pupillit liikkuvat luomien alla, mutta ei avaa silmiään.)*	Headings are used in all daily narratives. In Swedish daily narratives, the structure of headings seems to be obligatory. The headings are used typically as subjects.	*Circulation: Stable with inotrop.* *(Cirkulation: Stabil med inotropi.)* *Reacts only for pain stimulation during the suction of intubation tube.* *(Reagerar enbart vid smärtstimuli vid sugning i tuben.)*
Present and past participles are typical but verbs of *be*, *is* and *are* are not used.	*Consciousness remained unchanged.* *(Tajunta pysynyt ennallaan.)* *Blood pressure low.* *(Verenpaine matala.)*	Present and past participles are typical but verbs of *be*, *is* and *are* are not used.	*Breathing: Ventilator parameters unchanged.* *(Andning: Ventilator parametrarna oförändrade.)*
Complete sentences are rarely used.	*No spontaneous movements, rigidifies.* *(Ei spontaania liikettä, jäykistelee.)* *Husband and daughter have been staying a long time beside the patient.* *(Mies ja tytär olleet pitkään potilaan vierellä.)*	Complete sentences are rarely used.	*Light sedation, looks up now and then.* *(Lätt sederad, tittar upp ibland.)* *She took the wedding ring and the watch home.* *(Hon tog med sig vigselring och klocka hem.)*
Misspellings are found but the content or meaning is still clear.	*Hemodynamic – hemodynamic* *(Hemodynamiikka – henodynamiikka)*	Misspellings are found but the content or meaning is still clear.	*The motther is informed.* *(Mammman är informered.)* *Magnesium is addded.* *(Magnesium har tilllsatts.)*
Content	**Examples**	**Content**	**Examples**
The word *patient* as a subject or object is infrequently mentioned. If this word is mentioned it is not abbreviated.	*Oxidates well or ventilates well.* *(Happeutuu hyvin tai ventiloituu hyvin.)*	The word *patient* is used more often than in Finnish narratives as a subject or object. It is also replaced with abbreviations of *Pat* or *Pt.*	*Patient got a percutanous tracheostomy today.* *(Patienten har fått en perkutan trakeostomi idag.)* *Very worried about patient's condition.* *(Mycket oroliga över patientens tillstånd.)* *Pt. wakes up for talking and appears to be adequate.* *(Pt. vakner på tilltal och upplevs som adekvat.)*
Signs are typically used: e.g., >, <, -->, +, -.	*The height for the drain raised from 10 --> 20 mmHg.* *(Dreneerausrajaa nostettu 10 --> 20 mmHg.)* *Got medicine --> good response.* *(Sai lääkettä -->hyvä vaste.)*	Many different abbreviations are used. The origin of entire word is Swedish, English, Latin, professional or ICU typical.	*em.* [*eftermiddag, afternoon*], *HR* [*heart rate*], *VF* [*Ventricula/Fibrillation, Ventricular Fibrillation*]

Table 1. Special structural and contextual features of Finnish and Swedish daily ICU nursing narratives. The original examples are added in ().

I very rarely. If the reader is not a health care professional, a risk for confusing the subject (i.e., the patient or nurse) arises. However, the context makes it almost always clear who is referred to. Approximately half of the narratives do not contain any verb. The most common tense is perfect, but without the auxiliary *has*. When the meaning does not contain a subject it becomes "unnatural" to use *has*. Instead, the supine form is used, for example *slept*, *lain*, and *eaten*. Both present and past participles without *be*-verb are common, for example, *Breathing: Ventilator parameters unchanged.*

The use of headings is frequent and good – most of the time the content matches the headings (Tables 1 and 3). In addition, headings are used similarly in the Swedish and Finnish documents. Most of the time the headings are considered as subjects of the sentence, for example, *Consciousness: Unchanged. Liquor brighter than yesterday*.

However, in the use of headings there are two interesting findings: If the headings are to be chosen freely, as in the Finnish narratives, nurses tend to use their own headings and hence many synonyms or closely related concepts are used; for example, *hemodynamics* versus *blood pressure and pulse* or *breathing* versus *oxidation*. If the headings are obligatory, as in the Swedish narratives, nurses tend to write their observations under the heading which is somehow closest to the subject; for example, *body temperature* under *circulation* or *level of sedation* under *sleep*.

For both languages the use of different abbreviations is very common. Almost every daily nursing narrative included several abbreviations. Most of the abbreviations are typical for an ICU domain: *CVP* [*central venous pressure*], *PEEP* [*Positive End-Expiratory Pressure*], *EN* [*Enteral Nutrition*], *TPN* [*Total Parenteral Nutrition*], *pO2* [*partial pressure of oxygen*], *pCO2* [*partial pressure of carbon dioxide*], *MV* [*Minute Ventilation*] and *MAP* [*Mean Arterial Pressure*]. From a language technology point of view this means that ICU nursing narratives contain language-independent vocabulary. However, nurses in both countries also use many language dependent abbreviations.

3.2 Quantitative analysis

The Finnish data set (n=514) was quantitatively analyzed using the morphological analyser FinTWOL and the disambiguator FinCG, (Lingsoft 2010), and the Swedish data set (n=379) using the GTA, Granska Text Analyzer (Knutsson et al., 2003). Both data sets are rich in terms of amount of text and vocabulary (Table 2). It is also clear that the amount of text written per day and patient varies a lot in both data sets. More complex words were spelled in numerous ways. For example, the pharmacological substance Noradrenalin had approximately 350 and 60 different spellings in the Finnish and Swedish data sets, respectively. This problem is part of a more general issue of reference resolution e.g. when mapping different lexical terms referring to the same concept.

In our quantitative analysis, we have included punctuation characters. In the Swedish data there was a large amount of html-tags and other formatting characters, which has a high impact on the total number of tokens (see Table 2). Moreover, as Finnish is highly inflective, FinCG produces alternative lemmas, hence it is possible to reduce the sparseness of the data by processing the output by choosing only one alternative lemma (see total number of types in Table 2).

To further illustrate the richness of ICU nursing language, the number of unique bigrams (e.g., "*is not*", "*oxidate well*" and "*night time*" (note: a misspelled compound) are the most common ones for Finnish) and trigrams (e.g., "*oxygeneated and ventilated*", "*and ventilate well*" are among the most common ones for Finnish) were 368,166 (275,205 after FinCG) and 745,407 (356,307 after FinCG) for Finnish patient records. For the Swedish data, the number of unique bigrams was 469,455 (344,127 after GTA) and 1,064,944 (905,539 after GTA). Examples of common Swedish ICU bigrams and trigrams include "*circulation stabile*", "*during night*", "*in connection with*", and "*with good effect*". Of the content of Finnish nursing narratives, 11% are verbs, 7% nouns and less than 1% pronouns. For Swedish nursing narratives, the respective percentages are 11%, 27% and 2%. One reason for the high numbers for nouns in the Swedish data might be due to the large amount of (obligatory) headings relative to the Finnish data (see Table 3).

To support fluent information flow, language technology is needed to strengthen referential congruence. Much of this richness of vocabulary is explained by abbreviations and personal differences in professional jargon. In particular, abbreviations were common. Based on the analysis of the most common words, abbreviations were relatively established in Swedish data. For the Finnish data, abbreviations were less standard but *RR*, *SR*, *CVP*, *h*, *ad*, *ml*, *ok*, *vas.* [*vasen*, *left*] and *oik.* [*oikea*, *right*] were extremely common. Thus, referential congruence can be strengthened by spelling out the most common abbreviations automatically.

Adding topical content headings is another way to support information flow. Topical content headings were mandatory for Swedish data, but no default headings for Finnish existed. However, the headings for Finnish were established in terms of content. In Table 3, we see that the headings for both languages cover similar topics, which indicates that the clinical information need is similar for professionals in both countries (and languages). Thus, we recommend forming a standardized set of headings from which the user can voluntarily select the ones to be used. This does not exclude adding other headings. Another alternative is to develop language technology for topic segmentation and labeling. We have promising results from this approach (see, e.g., Suominen 2009).

Temporal expressions (e.g., *time*, *evening*, *night*) were often used in both data sets. This poses the question of tense analysis of verbs being unnecessary and the time-related words being enough to imply the needed temporal information. It is also interesting to note that the negations *inte* [*not*, Swe], *ingen* [*none*, Swe], *ej* [*not*, Swe] and *ei* [*no*/*not*, Fin] are all among the most common types, which is an important property to take into account in information extraction applications. Furthermore, words regarding the oral cavity, such as *breathing* and *mucus*, as well as relations, such as *daughter*, *son*, *wife*, and *husband* are very common in both data sets.

Inspired by the *tf*×*idf*-measure from information retrieval, we also analyzed the most common words in terms of a) the number of patients in whose documents the word was used and b) the number of daily nursing narratives in which the word was used. Here, we found, in both data sets, that those words that were used for all patients as well as all daily narratives, were very similar in both data sets, and were related to the most common headings, temporal expressions, negations and monitoring (e.g., *increase*, *continue*, *begin*).

The amount of Protected Health Information (PHI) in form of person names was equal in both of the data sets: 1.5 person names per thousand tokens. This is notable, since this has implications when it comes to integrity issues and reuse of data for research purposes.

FinCG did not recognize 36% of the content of Finnish nursing narratives. However, words marked as unrecognized by FinCG also included punctuation marks. In our previous study (see Suominen 2009 and references therein), we tailored FinCG by extending approximately 35,000 clinical terms. The extension not only substantially improved the applicability of FinCG to the health domain but also initiated piloting of our language technology components in an authentic healthcare environment in the fall 2008. This lead to the release of commercial language technology for Finnish health records (Lingsoft 2010).

Data	Finnish	Swedish
Total number of patients	514	379
Total number of tokens, types (unique tokens) and types after processing	 1,227,909 63,328 38,649	 1,959,271 - 41,883
Number of tokens per patient: Minimum Maximum Average Standard deviation	 540 14,118 2,389 1,635	 92 36,830 5,169 5,271
Total number of daily documents and shifts	 5,915 17,103	 4,700 –
Number of tokens per daily document: Minimum Maximum Average Standard deviation	 0 915 208 87	 5 9,389 417 239

Table 2. Comparison of Finnish and Swedish ICU data sets: total amount of text per patient. A daily document, i.e. nursing narrative, contains all text written about a given patient during a calendar day.

Finnish	n ≈	Swedish	n =
Hemodynamics	7,800	Respiratory	11,301
Consciousness	6,900	Circulation	10,630
Relatives	5,700	Elimination	10,041
Diuresis	5,400	Nutrition	8,258
Breathing	4,500	Communication	5,880
Oxygenation	3,600	Event Time	5,681
Other	3,200	Pain	4,732
Excretion	590	Psychosocial	4,682
Hemodialysis	370	Sleep	4,438
Pulse	160	Skin	4,402
Skin	160	Activity	3,794

Table 3. Comparison of Finnish and Swedish ICU data sets: the most common headings. For the Finnish data, where default headings were not given, we approximated the amount of heading by using an automated heuristics followed by manual combination of headings with the same meaning.

For Swedish, GTA handles unknown words differently than FinCG. However, by comparing the ICU words with a Swedish general language corpus (PAROLE, Gellerstam et al. (2000)), we found that 69% of the types are not included in PAROLE, which indicates a need for tailoring GTA (or similar tools for Swedish) with domain-specific ICU terms.

4 Conclusions

The purpose of this study was to do content and lexical analysis of nursing narratives written in an ICU. Our findings are that, even though the Finnish and Swedish languages are not linguistically closely related, the way of writing clinical nursing ICU narratives in both countries is very similar. Moreover, the written context made sentences clear for content experts, even though the texts were full of specialized jargon, misspellings, abbreviations, and missing subjects and objects. However, these characteristics make clinical text challenging for language technology. For example coreference resolution as in the case of noradrenalin.

We have also shown that the content characteristics of Finnish and Swedish ICU nursing narratives are very similar. This implies that developing tools for documentation support in ICUs is not country or language dependant in that respect. Developing such tools may improve possibilities for information extraction and text mining, enabling the possibilities to reuse the vast amounts of important practice-based information and evidence captured in clinical narratives. The framework we have introduced here could easily be employed in other studies of clinical texts.

6 Future work

In the future, we will use the results of this study in developing language technology for Finnish, Swedish and other Nordic ICU narratives. We will study how to identify abbreviations, misspellings and normalize and correct them, by using various distance measures and concept management techniques. We will also study how to automatically identify important parts of text and highlight them. Furthermore, we are interested in studying text provenance and pragmatics in this particular setting. In addition, we will evaluate the influence of these technology components in clinical practice. We will also address similarities and differences in clinical text written by various professional groups or at other hospital wards and health care units. Finally, we are eager to seek possibilities to incorporate laymen's information needs and their interaction with health care providers into our study.

Acknowledgments

We would like to thank Nordforsk and the Nordic Council of Ministers for the funding of our research network HEXAnord – HEalth teXt Analysis network in the Nordic and Baltic countries and NICTA, funded by the Australian Government as represented by the Department of Broadband, Communications and the Digital Economy and the Australian Research Council through the ICT Centre of Excellence program. We would also like to thank the Department of Information Technology and TUCS, University of Turku, Finland.

References

Lars Borin, Natalia Grabar, Catalina Hallett, Davis Hardcastle, Maria Toporowska Gronostaj, Dimitrios Kokkinakis, Sandra Williams and Alistair Willis. 2007. Empowering the patient with language technology. SemanticMining NoE 507505: *Deliverable D27.2.* <http://gup.ub.gu.se/gup/record/index.xsql?pubid=53590>

Grace Yuet-Chee Chung. 2009. Towards identifying intervention arms in randomized controlled trials: extracting coordinating constructions. *Journal of Biomedical Informatics.* 42(5):790–800

Hercules Dalianis, Martin Hassel and Sumithra Velupillai. 2009. The Stockholm EPR Corpus - Characteristics and Some Initial Findings. *Proceedings of ISHIMR 2009*, Evaluation and implementation of e-health and health information initiatives: international perspectives. 14th International Symposium for Health Information Management Research, Kalmar, Sweden, 14-16 October, 2009, pp 243-249, pdf. Awarded best paper.

Anna Ehrenberg, Margareta Ehnfors and Ingrid Thorell-Ekstrand. 1996. Nursing documentation in patient records: experience of the use of the VIPS model. *Journal of Advanced Nursing* 24, 853–867.

Martin Gellerstam, Yvonne Cederholm, and Torgny Rasmark. The bank of Swedish. In: *Proceedings of LREC 2000 -- The 2nd International Conference on Language Resources and Evaluation*, pages 329–333, Athens, Greece.

Udo Hahn and Joachim Wermter. 2004. High-performance tagging on medical texts. *Proceedings of the 20th international conference on Computational Linguistics*. Geneva, Switzerland.

Henk Harkema, Dowling JN, Thornblade T, Chapman WW. 2009. ConText: an algorithm for determining negation, experiencer, and temporal status from clinical reports. *Journal of Biomedical Informatics* 2009;42(5):839–51.

Ragnhild Hellesø. 2005. Information handling in the nursing discharge notes, *Journal of Clinical Nursing*, Volume 15 Issue 1, 11 - 21. Blackwell publishing

Gunlög Josefsson. 1999. Få feber eller tempa? Några tankar om agentivitet i medicinskt fackspråk, *Alla tiders språk: en vänskrift till Gertrud Pettersson*. Pages 127. Institutionen för nordiska språk. Lund. (In Swedish)

Hyeoneui Kim, Sergey Goryachev, Craciela Rosemblat, Allen Browne, Alla Keselman and Qing Zeng-Treitler. 2007. Beyond surface characteristics: a new health text-specific readability measurement. *AMIA Annual Symp*. 11:418-22.

Ola Knutsson, Johnny Bigert, and Vigg Kann. 2003. A robust shallow parser for Swedish. *In Proceedings 14th Nordic Conf. on Comp. Ling. NODALIDA*.

Sirkka Lauri and Sanna Salanterä;. 2002. Developing an instrument to measure and describe clinical decision making in different nursing fields. *Journal of Professional Nursing*. Mar-Apr;18(2), 93-100.

Lingsoft. 2010, Lingsoft Oy, http://www.lingsoft.fi/

Clement J. McDonald. 1997. The Barriers to Electronic Medical Record Systems and How to Overcome Them. *JAMIA*. 1997;4:213–221.

Stéphane M. Meystre, Guergana K. Savova, Karin C. Kipper-Schuler and John E. Hurdle. 2008. Extracting Information from Textual Documents in the Electronic Health Record: a Review of Recent Research. *Yearbook Med Inform*. 2008:128-44.

Raymond L. Ownby 2005. Influence of Vocabulary and Sentence Complexity and Passive Voice on the Readability of Consumer-Oriented Mental Health Information on the Internet. *AMIA Annual Symposium Proceedings*. 2005: 585–588.

Serguei V. S. Pakhomov, Anni Coden and Christopher G. Chute. 2006. Developing a corpus of clinical notes manually annotated for part-of-speech. *International Journal of Medical Informatics*. 2006 Jun;75(6):418-29. Epub 2005 Sep 19.

Patientdatalagen (2008:355) *Svensk författningssamling*, Socialdepartementet, 2008, Stockholm. (In Swedish)

Hanna Suominen. 2009. Machine Learning and Clinical Text: Supporting Health Information Flow. *TUCS Dissertations* No 125, Turku Centre for Computer Science, 2009, Turku, Finland.

Hanna Suominen, Heljä Lundgrén-Laine, Sanna Salanterä, Helena Karsten, and Tapio Salakoski. 2009. Information flow in intensive care narratives. In Chen J, Chen C, Ely J, Hakkani-Tr D, He J, Hsu H.-H, Liao L, Liu C, Pop M, Ranganathan S, Reddy C.K, Ruan J, Song Y, Tseng V.S, Ungar L, Wu D, Wu Z, Xu K, Yu H, Zelikovsky A, editors. *Proceedings IEEE International Conference on Bioinformatics and Biomedicine Workshops*, BIBM 2009, pages 325–330. Institute of Electrical and Electronics Engineers, Los Alamitos, California, USA.

Kaarina Tanttu and Helena Ikonen. 2007. Nationally standardized electronic nursing documentation in Finland by the year 2007. *Stud Health Technol Inform*.122:540-1.

Asta Thoroddsen, Kaija Saranto, Anna Ehrenberg, Walter Sermeus. 2009. Models, standards and structures of nursing documentation in European countries. *Stud Health Technol Inform*.146:327-31.

Katrin Tomanek, Joachim Wermter and Udo Hahn. 2007. A Reappraisal of sentence and token splitting for life sciences documents. *Stud Health Technol Inform*. 129 (Pt 1):524-8.

Webster's 2010. Webster's New World Medical Dictionary. http://www.medterms.com/script/main/hp.asp,last visited February 2, 2010.

Extracting Medication Information from Discharge Summaries

Scott Halgrim, Fei Xia, Imre Solti, Eithon Cadag
University of Washington
PO Box 543450
Seattle, WA 98195, USA
{captnpi,fxia,solti,ecadag}@uw.edu

Özlem Uzuner
University of Albany, SUNY
135 Western Ave
Albany, NY 12222, USA
ouzuner@uamail.albany.edu

Abstract

Extracting medication information from clinical records has many potential applications and was the focus of the i2b2 challenge in 2009. We present a hybrid system, comprised of machine learning and rule-based modules, for medication information extraction. With only a handful of template-filling rules, the system's core is a cascade of statistical classifiers for field detection. It achieved good performance that was comparable to the top systems in the i2b2 challenge, demonstrating that a heavily statistical approach can perform as well or better than systems with many sophisticated rules. The system can easily incorporate additional resources such as medication name lists to further improve performance.

1 Introduction

Narrative clinical records store patient medical information, and extracting this information from these narratives supports data management and enables many applications (Levin et al., 2007). Informatics for Integrating the Biology and the Bedside (*i2b2*) is an NIH-funded National Center for Biomedical Computing based at Partners HealthCare System, and it has organized annual NLP shared tasks and challenges since 2006 (https://www.i2b2.org/). The Third i2b2 Workshop on NLP Challenges for Clinical Records in 2009 studied the extraction of medication information from hospital discharge summaries (https://www.i2b2.org/NLP/Medication/), a task we refer to as *the i2b2 challenge* in this paper.

In the past decade, there has been extensive research on information extraction in both the general and biomedical domains (Wellner et al., 2004; Grenager et al., 2005; Poon and Domingos, 2007; Meystre et al, 2008; Rozenfeld and Feldman, 2008). Interestingly, despite the recent prevalence of statistical approaches in most NLP tasks (including information extraction), most of the systems developed for the i2b2 challenge were rule-based. In this paper we present our hybrid system, whose core is a cascade of statistical classifiers that identify medication fields such as medication names and dosages. The fields are then assembled to form medication entries. While our system did not participate in the i2b2 challenge (as we were part of the organizing team), it achieved good results that matched the top i2b2 systems.

2 The i2b2 Challenge

This section provides a brief introduction to the i2b2 challenge.

2.1 The task

The i2b2 challenge studied the automatic extraction of information corresponding to the following *fields* from hospital discharge summaries (Uzuner, et al., 2010a): names of medications (*m*) taken by the patient, dosages (*do*), modes (*mo*), frequencies (*f*), durations (*du*), and reasons (*r*) for taking these medications. We refer to the medication field as the *name* field and the other five fields as the *non-name* fields. All non-name fields correspond to some name field mention; if they were specified within a two-line window of that name mention,

Proceedings of the NAACL HLT 2010 Second Louhi Workshop on Text and Data Mining of Health Documents, pages 61–67,
Los Angeles, California, June 2010. ©2010 Association for Computational Linguistics

the i2b2 challenge required such fields to be linked to the name field to form an *entry*. For each entry, a system must determine whether the entry appeared in a list of medications or in narrative text. Table 1 shows an excerpt from a discharge summary and the corresponding entries in the gold standard. The first entry appears in narrative text, and the second in a list of medication information.

Excerpt of Discharge Summary
55 the patient noted that he had a recurrence of this 56 vague chest discomfort as he was sitting and 57 talking to friends. He took a sublingual 58 Nitroglycerin without relief. ... 65 Flomax (Tamsulosin) 0.4 mg, po, qd,...
Gold standard: m="Nitroglycerin" 58:0 58:0 ||do="nm"|| mo="sublingual" 57:6 57:6 ||f="nm" ||du="nm" || r="vague chest discomfort" 56:0 56:2 || ln="narrative" ... m="flomax (tamsulosin)" 65:0 65:3||do="0.4 mg" 65:4 65:5||mo="po" 65:6 65:6||f="qd" 65:7 65:7||du="nm"||r="nm"||ln="list"

Table 1: A sample discharge summary excerpt and the corresponding entries in the gold standard. The fields inside an entry are separated by "||". Each field is represented by the string and its position (i.e., "line number: token number" offsets). "nm" means the field value is not mentioned for this medication name.

2.2 Data Sets

The i2b2 challenge used a total of 1243 discharge summaries:

- 696 of these summaries were released to participants for system development, and the i2b2 organizing team provided the gold standard annotation for 17 of them.

- Participating teams could choose to annotate more files themselves. The University of Sydney team annotated 145 out of the 696 summaries (including re-annotating 14 of the 17 files annotated by the i2b2 organizing team) and generously shared their annotations with i2b2 after the challenge for future research. We obtained and used 110 of their annotations as our training set and the remaining 35 summaries as our development set.

- The participating teams produced system outputs for 547 discharge summaries set aside for testing. After the challenge, 251 of these summaries were annotated by the challenge participants, and these 251 summaries formed the final test set (Uzuner et al., 2010b).

The sizes of the data sets used in our experiments are shown in Table 2. The training and development sets were created by the University of Sydney, and the test data is the i2b2 official challenge test set. The average number of entries and fields vary among the three sets because the summaries in the test set were chosen *randomly* from the 547 summaries, whereas the University of Sydney team annotated the *longest* summaries.

2.3 Additional resources

Besides the training data, the participating teams were allowed to use any additional tools and resources that they had access to, including resources not available to the public. All challenge participants used additional resources such as UMLS (www.nlm.nih.gov/research/umls/), but the exact resources used varied from team to team. Therefore, the challenge was similar to the so-called *open-track* challenge in the general NLP field, as opposed to a *closed-track* challenge that could require that all the participants use only the list of resources specified by the challenge organizers

2.4 Evaluation metrics

The i2b2 challenge used two sets of evaluation metrics: *horizontal* and *vertical* metrics. Horizontal metrics measured system performance at the entry level, whereas vertical metrics measured system performance at the field level. Both sets of metrics compared the system output and the gold standard at the *span* level for *exact match* and at the *token* level for *inexact match*, using precision, recall, and F-score (Uzuner et al., 2010a). The primary metric for the challenge is exact horizontal F-score, which is the metric we use to evaluate our system.

Data Sets	# of Summaries	# of Entries	# of Fields	# of Names	# of Doses	# of Freq	# of Modes	# of Duration	# of Reason
Training set	110	5970 (54.3)	14886 (135.3)	5684 (51.7)	2929 (26.6)	2740 (24.9)	2146 (19.5)	302 (2.7)	1085 (9.9)
Dev set	35	2401 (68.6)	5988 (171.1)	2302 (65.8)	1163 (33.2)	1096 (31.3)	880 (25.1)	111 (3.2)	436 (12.5)
Test set	251	8936 (35.6)	22041 (87.8)	8495 (33.8)	4387 (17.5)	3999 (15.9)	3307 (13.2)	511 (2.0)	1342 (5.3)

Table 2: The data sets used in our experiments. The numbers in parentheses are the average numbers of entries or fields per discharge summary.

2.5 Participating systems

Twenty teams participated in the challenge. Fifteen teams used rule-based approaches, and the rest used statistical or hybrid approaches. The performances of the top five systems are shown in Table 3. Among them, only the top system, developed by the University of Sydney, used a hybrid approach, whereas the rest were rule-based.

Rank	Precision	Recall	F-score
1	89.6	82.0	85.7
2	84.0	80.3	82.1
3	86.4	76.6	81.2
4	78.4	82.3	80.3
5	84.1	75.8	79.7

Table 3: The performance (exact horizontal precision/recall/F-score) of the top five i2b2 systems on the test set.

3 System description

We developed a hybrid system with three processing steps: (1) a preprocessing step that finds section boundaries, (2) a field detection step that identifies the six fields, and (3) a field linking step that links fields together to form entries. The second step is a statistical system, whereas the other two steps are rule-based. The second step was the main focus of this study.

3.1 Preprocessing

In addition to common processing steps such as part-of-speech (POS) tagging, our preprocessing step includes a section segmenter that breaks discharge summaries into sections. Discharge summaries tend to consist of sections such as 'ADMIT DIAGNOSIS', 'PAST MEDICAL HISTORY', and 'DISCHARGE MEDICATIONS'. Knowing section boundaries is important for the i2b2 challenge because, according to the annotation guidelines for creating the gold standard, medications occurring under certain sections (e.g., family history and allergic reaction) should be excluded from the system output. Furthermore, knowing the types of sections could be useful for field detection and field linking; for example, entries in the 'DISCHARGE MEDICATIONS' section are more likely to appear in a list of medications than in narrative text.

The set of sections and the exact spelling of section headings vary across discharge summaries. The section segmenter uses regular expressions (e.g., '^\s*([A-Z\s]+):' -- a line starting with a sequence of capitalized words followed by a colon) to collect potential section headings from the training data, and the headings whose frequencies are higher than a threshold are used to identify section boundaries in the discharge summaries.

3.2 Field detection

This step consists of three modules: the first module, *find_name*, finds medication names in a discharge summary; the second module, *context_type*, processes each medication name identified by *find_name* and determines whether the medication appears in narrative text or in a list of medications; the third module, *find_others*, detects the five non-name field types.

For *find_name* and *find_others*, we follow the common practice of treating named-entity (NE) detection as a sequence labeling task with the BIO tagging scheme; that is, each token in the input is tagged with B-x (beginning an NE of type x), I-x (inside an NE of type x) and O (outside any NE).

3.2.1 The *find_name* module

As this module identifies medication names only, the tagset under the BIO scheme has three tags: B-m for beginning of a name, I-m for inside a name, and O for outside. Various features are used for this module, which we group into four types:

- (F1) includes word n-gram features (n=1,2,3). For instance, the bigram w_{i-1} w_i looks at the current word and the previous word.
- (F2) contains features that check properties of the current word and its neighbors (e.g., the POS tag, the affixes and the length of a word, the type of section that a word appears in, whether a word is capitalized, whether a word is a number, etc.)
- (F3) checks the BIO tags of previous words
- (F4) contains features that check whether n-grams formed by neighboring words appear as part of medication names in given medication name lists. The name lists can come from labeled training data or additional resources such as UMLS.

3.2.2 The *context_type* module

This module is a binary classifier which determines whether a medication name occurs in a list or narrative context. Features used by this module include the section name as identified by the preprocessing step, the number of commas and words on the current line, the position of the medication name on the current line, and the current and nearby words.

3.2.3 The *find_others* module

This module complements the *find_name* module and uses eleven BIO tags to identify five non-name fields. The feature set used in this module is very similar to the one used in *find_name* except that some features in (F2) and (F4) are modified to suit the needs of the non-name fields. For instance, a feature will check whether a word fits a common pattern for dosage. In addition, some features in (F2) look at the output of previous modules: e.g., the location of nearby medication names as this information can be provided by the *find_name* module at test time.

3.3 Field linking

Once medication names and other fields have been found, the final step is to form entries by associating each medication name with its related fields. Our current implementation uses simple heuristics. First, we go over each non-name field and link it with the closest *preceding* medication name unless the distance between the non-name field and its closest *following* medication name is much shorter. Second, we assemble the (name, non-name) pairs to form medication entries with a few rules.

More information about the modules discussed in this section and features used by the modules is available in (Halgrim, 2009).

4 Experimental results

In this section, we report the performance of our system on the development set (Section 4.1-4.3) and the test set (Section 4.4). The data sets are described in Table 2. For all the experiments in this section, unless specified otherwise, we report exact horizontal precision/recall/F-score, the primary metrics for the i2b2 challenge.

For the three modules in the field detection step, we use the Maximum Entropy (MaxEnt) learner in the Mallet package (McCallum, 2002) because, in general, MaxEnt produces good results without much parameter tuning and the training time for MaxEnt is much faster than more sophisticated algorithms such as CRF (Lafferty et al., 2001).

To determine whether the difference between two systems' performances is statistically significant, we use approximate randomization tests (Noreen, 1989) as follows. Given two systems that we would like to compare, we first calculate the difference between exact horizontal F-scores. Then two pseudo-system outputs are generated by randomly swapping (at 0.5 probability) the two system outputs for each discharge summary. If the difference between F-scores of these pseudo-outputs is no less than the original F-score difference, a counter, *cnt,* is increased by one. This process was repeated n=10,000 times, and the p-value of the significance is equal to $(cnt+1)/(n+1)$.

If the p-value is smaller than a predefined threshold (e.g., 0.05), we conclude that the difference between the two systems is statistically significant.

4.1 Performance of the whole system

4.1.1 Effect of feature sets

To test the effect of feature sets on system performance, we trained *find_name* and *find_others* with different feature sets and tested the whole system on the development set. For (F4), we used two medication name lists. The first list consists of medication names gathered from the training data. The second list includes drug names from the FDA database (www.accessdata.fda.gov/scripts/cder/ndc/). We use the second list to test whether adding features that check the information in an additional resource could improve the system performance.

The results are in Table 4. For the last two rows, F1-F4a uses the first medication name list, and F1-F4b uses both lists. The F-score difference between all adjacent rows is statistically significant at p≤0.01, except for the pair F1-F3 vs. F1-F4a. It is not surprising that using the first medication name list on top of F1-F3 does not improve the performance, as the same kind of information has already been captured by F1 features. The improvement of F1-F4b over F1-F4a shows that the system can easily incorporate additional resources and achieve a statistically significant (at p≤0.01) gain.

Features	Precision	Recall	F-score
F1	72.5	60.3	65.8
F1-F2	82.5	78.2	80.3
F1-F3	88.4	77.9	82.8
F1-F4a	87.4	77.9	82.4
F1-F4b	88.1	79.4	83.5

Table 4: System performance on the development set with different feature sets

4.1.2 Effect of training data size

Figure 1 shows the system performance on the development set when different portions of the training set are used for training. The curve with "+" signs represents the results for F1-F4b, and the curve with circles represents the results for F1-F4a. The figure illustrates that, as the training data size increases, the F-score with both feature sets improves. In addition, the additional resource is most helpful when the training data size is small, as indicated by the decreasing gap between the two sets of F-scores when the size of training data increases.

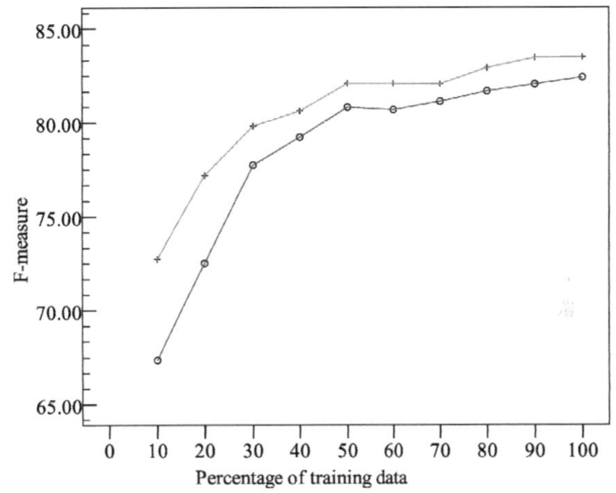

Figure 1: System performance on the development set with different training data sizes (Legend: ○ represents F-scores with features in F1-F4a; + represents F-scores with features in F1-F4b)

4.1.3 Pipeline vs. *find_all*

The current field detection step is a pipeline approach with three modules: *find_name*, *context_type*, and *find_others*. Having three separate modules allows each module to choose the features that are most appropriate for it. In addition, later modules can use features that check the output of the previous modules. A potential downside of the pipeline system is that the errors in the early module would propagate to later modules. An alternative is to use a single module to detect all six field types together.

Figure 2 shows the result of *find_all* in comparison to the result for the three-module pipeline. Both use the F1-F4b feature sets, except that *find_others* uses some features that check the output of previous modules which are not available to *find_all*.

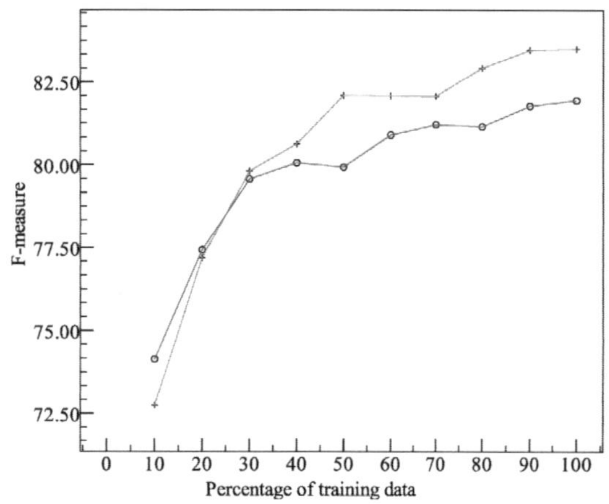

Figure 2: Pipeline vs. *find_all* for field detection (Legend: ○ represents F-scores with *find_all*; + represents F-scores with the three-module pipeline)

Interestingly, when 10% of the training set is used for training, *find_all* has a higher F-score than the pipeline approach, although the difference is not statistically significant at $p \leq 0.05$. As more data is used for training, the pipeline outperforms *find_all*, and when at least 50% of the training data is used, the difference between the two is statistically significant at $p \leq 0.05$. One possible explanation for this phenomenon is that as more training data becomes available, the early modules in the pipeline make fewer errors; as a result, the disadvantage of the pipeline approach caused by error propagation is outweighed by the advantage that the later modules in the pipeline can use features that check the output of the earlier modules.

4.2 Performance of the field detection step

Table 5 shows the *exact* precision/recall/F-score on identifying the six field types, using all the training data, F1-F4b features, and the pipeline approach for field detection. A span in the system output *exactly matches* a span in the gold standard if the two spans are identical and have the same field type. Among the six fields, the results for duration and reason are the lowest. That is because duration and reason are longer phrases than the other four field types and there are fewer strong, reliable cues to signal their presence.

When making the narrative/list distinction, the accuracy of our *context_type* module is 95.4%. In contrast, the accuracy of the baseline (which treats each medication name as in a list context) is only 55.6%.

	Precision	Recall	F-score
Name	91.2	88.5	89.9
Dosage	96.6	90.8	93.6
Frequency	93.9	89.0	91.8
Mode	95.7	90.3	92.9
Duration	73.8	43.2	54.5
Reason	72.2	31.0	43.3
All fields	92.6	84.5	88.4

Table 5: The performance (exact precision/recall/F-score) of field detection on the development set.

4.3 Performance of the field linking step

In order to evaluate the field linking step, we generated a list of (name, non-name) pairs from the gold standard, where the name and non-name fields appear in the same entry in the gold standard. We then compared these pairs with the ones produced by the field linking step and calculated precision/recall/F-score. Table 6 shows the result of two experiments: in the cheating experiment, the input to the field linking step is the fields from the gold standard; in the non-cheating experiment, the input is the output of the field detection step. These experiments show that, while the heuristic rules used in this step work reasonably well when the input is accurate, the performance deteriorates considerably when the input is noisy, an issue we plan to address in future work.

	Precision	Recall	F-score
Non-cheating	87.4	75.1	80.8
Cheating	96.2	94.5	95.3

Table 6: The performance of the field linking step on the development set (cheating: assuming perfect field input; non-cheating: using the output of the field detection step)

4.4 Results on the test data

Table 7 shows the system performance on the i2b2 official test data. The system was trained on the union of the training and development data. Compared with the top five i2b2 systems (see Table 3), our system was second only to the best i2b2 system, which used more resources and more sophisticated rules for field linking (Patrick and Li, 2009).

	Precision	Recall	F-score
Horizontal	**88.6**	**80.2**	**84.1**
Name	92.6	87.1	89.8
Dosage	96.3	90.2	93.1
Frequency	95.6	90.8	93.2
Mode	96.7	90.2	93.3
Duration	70.6	40.5	51.5
Reason	73.4	34.7	47.1
All fields	91.6	82.7	86.9

Table 7: System performance on the test set when trained on the union of the training and the development sets with F1-F4b features.

5 Conclusion

We present a hybrid system for medication extraction. The system is built around a pipeline of cascading statistical classifiers for field detection. It achieves good performance that is comparable to the top systems in the i2b2 challenge, and incorporating additional resources as features further improves the performance. In the future, we plan to replace the current rule-based field linking module with a statistical module to improve accuracy.

Acknowledgments

This work was supported in part by US DOD grant N00244-091-0081 and NIH Grants 1K99LM010227-0110, U54LM008748, and T15LM007442-06. We would also like to thank three anonymous reviewers for helpful comments.

References

Trond Grenager, Dan Klein, and Christopher Manning. 2005. Unsupervised learning of field segmentation models for information extraction. In Proc. of ACL-2005.

Scott Halgrim. 2009. A Pipeline Machine Learning Approach to Biomedical Information Extraction. Master Thesis. University of Washington.

J. Lafferty and A. McCallum and F. Pereira. 2001. Conditional random fields: Probabilistic models for segmenting and labeling sequence data. In Proc. of the 18th International Conference on Machine Learning (ICML-2001).

Matthew A. Levin, Marina Krol, Ankur M. Doshi, and David L. Reich. 2007. Extraction and mapping of drug names from free text to a standardized nomenclature. *AMIA Symposium Proceedings,* pp 438-442.

S. Meystre, G. Savova, K. Kipper-Schuler, and J. Hurdle. 2008. Extracting Information from Textual Documents in the Electronic Health Record: A Review of Recent Research. *IMIA Yearbook of Medical Informatics Methods Inf Med 2008*; 47 Suppl 1:128-44.

Andrew McCallum. 2002. Mallet: A Machine Learning for Language Toolkit. http://mallet.cs.umass.edu

Eric W. Noreen. 1989. Computer intensive methods for testing hypotheses: an introduction. John Wiley & Sons.

Jon Patrick and Min Li, 2009. A Cascade Approach to Extract Medication Event (i2b2 challenge 2009). Presentation at the Third i2b2 Workshop, November 2009, San Francisco, CA.

Hoifung Poon and Pedro Domingos. 2007. Joint inference in information extraction. In Proc. of the National Conference on Artificial Intelligence (AAAI), pp 913-918.

Benjamin Rozenfeld and Ronen Feldman. 2008. Self-supervised relation extraction from the web. Knowledge and Information Systems, 17(1):17-33.

Özlem Uzuner, Imre Solti, and Eithon Cadag, 2010a. Extracting Medication Information from Clinical Text. Submitted to JAMIA.

Özlem Uzuner, Imre Solti, Fei Xia, and Eithon Cadag, 2010b. Community Annotation Experiment for Ground Truth Generation for the i2b2 Medication Challenge. Submitted to JAMIA.

Ben Wellner, Andrew McCallum, Fuchun Peng, and Michael Hay. 2004. An integrated, conditional model of information extraction and coreference with application to citation matching. In Proc. of the 20th Conference on Uncertainty in AI (UAI-2004).

Linking SweFN++ with Medical Resources, towards a MedFrameNet for Swedish

Dimitrios Kokkinakis
Department of Swedish Language, Språkbanken
University of Gothenburg
SE-405 30, Gothenburg, Sweden
dimitrios.kokkinakis@svenska.gu.se

Maria Toporowska Gronostaj
Department of Swedish Language, Språkbanken
University of Gothenburg
SE-405 30, Gothenburg, Sweden
maria.gronostaj@svenska.gu.se

Abstract

In this pilot study we define and apply a methodology for building an event extraction system for the Swedish scientific medical and clinical language. Our aim is to find and describe linguistic expressions which refer to medical events, such as events related to diseases, symptoms and drug effects. In order to achieve this goal we have initiated actions that aim to extend and refine parts of the ongoing compilation of the Swedish FrameNet++ (SFN++). SFN++, as its English original predecessor, is grounded in *Frame Semantics* which provides a sound theoretical ground for modeling and linking linguistic structures encountered in general language and in specific domains (after specialization). Using such resource we have started to manually annotate domain texts for enriching SFN++ with authentic samples and for providing training data for automated event extraction techniques.

1 Introduction

In the clinical setting vast amounts of health-related data are constantly collected, while medical and biomedical scientific publications, in e.g. molecular biology, genetics, proteomics and other types of -omics, increase in a dramatic manner. These data are undoubtfully a valuable source of evidence-based research. However, to empower researchers to make highly efficient use of the resulting volume of literature and the knowledge that is encoded therein, the material must be better integrated and linked via effective automated processing. Tools have to be developed for the automatic processing of the textual content in a deeper, more semantically-oriented fashion having access to multilayered lexical and grammatical information. The goal is then to enable rapid, effective and as far as possible accurate extraction of relationships, facts and events asserted and described in the data. Event extraction is understood here as an activity, that broadly follows the BioNLP 2009 shared task view (Kim *et al.*, 2009), in which an event is considered to be an involvement of multiple entities in varying roles. The task is fundamental to the objective of Language Technology systems, such as Question-Answering and Information Extraction (IE), which have as their higher-level goal to identify instances of a particular class of events (or relationships) in a text and to extract their relevant arguments. We argue that such information has a direct correlation with FrameNet's semantic frames, since templates in the context of IE are frame-like structures with slots representing the event basic information. Our intention is to explore the applicability of SFN++ to the clinical and scientific medical domain in Swedish. Therefore, relevant domain specific entities are explicitly annotated by automatic indexing of the texts by the Swedish and English Medical Subject Headings thesauri (MeSH); *cf.* Kokkinakis (2009). Non-medical entities such as temporal expressions, locative expressions and personal characteristics such as gender and age are provided by an extended named entity recognition process (Kokkinakis, 2004). The partial syntactic analysis that follows is aware of the preprocessing steps and uses the background knowledge as features encoded in XML using the TIGER-XML format (Brants & Hansen, 2002).

2 Background

Methods employed in the extraction of events have generally involved two approaches. The first oneis based on manual annotation of events in domain-specific text samples in order to create training

Proceedings of the NAACL HLT 2010 Second Louhi Workshop on Text and Data Mining of Health Documents, pages 68–71,
Los Angeles, California, June 2010.

resources for processes that may *learn* to recognize events in new texts (Kim *et al.*, 2008). The second is based on methods in which events are automatically acquired from unannotated texts; Nobata & Sekine (1999), in the sense that no manually pre-encoded training resources are used for producing the extraction patterns. Both methodologies have produced rapid advances in the robustness and applicability of IE. We believe that Frame Semantics (Fillmore, 1976) is a suitable resource, for the first type of method, and in our work we started to explore means for specializing and refining parts of the ongoing development of the SFN++ (Borin *et al.*, 2009), on the medical domain. Our goal is to enrich domain corpora with layers of syntactic and semantic information providing relevant support for IE and text mining research in the field.

2.1 FrameNet

FrameNet <http://framenet.icsi.berkeley.edu> is based on the principles of Frame Semantics supported by corpus evidence. A semantic frame is a script-like structure of concepts, which are linked to the meanings of linguistic units and associated with a specific event or state. Each frame identifies a set of frame elements, which are frame specific semantic roles (e.g. participants or arguments). FN documents the range of semantic and syntactic combinatory possibilities of frame evoking lexical units, phrases and clauses. FN facilitates modeling the mapping of form and meaning within these structures in the medical discourse through manual annotation of example sentences and automatic summarization of the resulting annotations. A word can evoke a frame, and its syntactic dependents can fill the frame element slots (Dolbey *et al.*, 2006). Moreover, since a single frame element can have different grammatical realizations it can enhance the investigation of combinatorial possibilities more precisely than other standard lexical resources such as WordNet (*cf.* Dolbey, 2009).

2.2 FrameNet and Information Extraction

IE is a technology that has a direct correlation with frame-like structures; since templates in the context of IE are frame-like structures with slots representing the event information. IE operates on specific domains, and the objective of IE systems is to identify instances of a particular class of events or relationships in a text and to extract the relevant arguments of the event or relationship. It has been stated Kilgarriff (1997) that the task requires the key terms of that domain, the "foreground lexicon", to be tightly bound to the domain vocabularies (e.g. ontology) and having well-articulated meaning representations. According to this philosophy the foreground lexicon for a domain will generally contain:

- the key predicates for the domain (*trigger words*);
- the sets of lexical items which realize the predicate (*lexical units*);
- how they and their arguments relate to the IE system's output formalism (*core elements and valencies*);
- how their complements relate to the predicate's arguments (*non-core elements*).

3 Methodology

A subset of the original English FN already contains relevant frames with direct applicability to the medical domain, such as: *Medical conditions*, *Experience bodily harm* and *Cure* (see the Appendix for the SFN++ Cure frame; a snapshot from: <http://spraakbanken.gu.se/swefn/>); in the figure 'saldo' is the name of a freely available modern Swedish semantic and morphological resource intended for language technology applications; *cf.* Borin & Forsberg, 2009). We keep the English labels for each frame, while the content is manually adapted to Swedish. We start by identifying words or phrases that evoke frames and assigning the correct frame to them interactively using the SALTO tool (Burchardt *et al.*, 2006). For each instance of a frame that has been found, we label the parts of the sentence that fill their semantic roles. Sentences that fulfill these criteria are selected from a large corpus after preprocessed by a number of pipelined tools including: multiword and idiom expression identification, part-of-speech tagging, lemmatization, named entity and terminology recognition, shallow parsing, using a cascaded parser (Abney, 1996) and XML formatting according to the TIGER-XML. A simplified example (i.e. *The doctor treated her with cortisone*) would at the end of the processing be labeled as:

FRAME	[H Läkaren]	[LU *behandlade*]	[P henne]	[M med kortison]
CURE	HEALER	LEXICAL UNIT	PATIENT	MEDICATION

4 Conclusions and Further Research

Our work is a first attempt to get a whole picture of the requirements and difficulties for specializing SweFN++ to a domain and gaining experience on applying it to a sublanguage. Our goal for the near future is to aid the development of a fully automated event extraction system for the Swedish medical domain. This can be accomplished by annotating various types of data, in the medical sublanguage, and classifying text segments to the class type of the event mentioned. Then, by applying other means (e.g. pattern matching rules) we can extract the participants of the events and match to e.g. information extraction templates, depending on different applications and needs. Of course, there are several other issues that need to be worked on, such as devising ways to recognize negated and/or speculative language usage. Mapping medical frame elements onto corresponding concepts in a thesaurus-based lexicon turns a relatively little lexical resource into a more robust one and hence more useful for semi-automatic semantic annotation of corpora; *cf.* Baker *et al.*, 2007. For annotating the Swedish corpus, we have used our intern thematically sorted lexicons with medical vocabulary and the Swedish data from MeSH.

Core FEs in FN	MESH thesaurus nodes
Ailment, Affliction	Disease
Body_parts	Anatomy
Medication	Chemicals and Drugs
Treatment	Analytical, Diagnostic and Therapeutic Techniques and Equipment
Patient	Person

Table 1. Example of mapping of core frame elements onto MeSH top nodes

The advantage of the pre-processing stage is very important and we believe that there is a feasible way to proceed in order to aid the annotation of large textual samples. Preliminary quantitative analysis of the examined instances has shown that many linguistically optional scheme elements need to be re-ranked whenever viewed from a medical pragmatic perspective. For example *Time*, *Measure* and *Method* provide relevant data for diagnosing patients' health condition. Another fact that might need special attention is the issue of tagging pronouns. It seems that these should not be tagged before anaphoric relations and their functional roles have been established. This is particularly important for distinguishing between patients and health care providers. Use of frame-semantic resources in general for language technology is evident. However, the effect of adding frame-semantic information to LT applications has been rather low, our work intends to change this situation in the near future, getting insights into the deeper semantics of the domain events.

References

Abney S. 1996. Partial Parsing via Finite-State Cascades. *J. Nat. Lang. Eng.*, **2**(4): 337-344.

Baker C., Ellsworth M. and Erk. K. 2007. *SemEval'07 task 19: frame semantic structure extraction.* Proceedings of the 4th International Workshop on Semantic Evaluations. Prague, Czech Republic

Borin L. and Forsberg M. 2009. *All in the family: A comparison of SALDO and WordNet.* Nodalida Workshop: WordNets and other Lexical Semantic Resources - between Lexical Semantics, Lexicography, Terminology and Formal Ontologies. Odense.

Borin L., Dannélls D., Forsberg M., Toporowska Gronostaj M. and Kokkinakis D. 2009. *Thinking Green: Toward Swedish FrameNet++.* FN Masterclass & Workshop. Milan, Italy.

Brants S. and Hansen S. 2002. *Developments in the TIGER Annotation Scheme and their Realization in the Corpus.* 3rd Language Resources and Evaluation (LREC). Pp. 1643-1649 Las Palmas.

Burchardt A., Erk K., Frank A., Kowalski A. and Pado S. 2006. *SALTO – A Versatile Multi-Level Annotation Tool.* 5th Language Resources and Evaluation (LREC). Genoa.

Dolbey A., Ellsworth M. and Scheffczyk J. 2006. *BioFrameNet: A Domain-specific FrameNet Extension with Links to Biomedical Ontologies.* KR-MED: Bio. Ontology in Action. Maryland, USA.

Dolbey A. 2009. *BioFrameNet, a FrameNet Extension to the Domain of Molecular Biology.* FRAMES AND CONSTRUCTIONS. A conference in honor of Charles J. Fillmore. U of California, Berkeley.

Fillmore C. J. 1976. *Frame semantics and the nature of language.* NY Academy of Sciences: Conference on the Origin and Development of Lang. & Speech, Vol. 280: 20-32.

Kilgarriff A. 1997. *Foreground and Background Lexicons and Word Sense Disambiguation for Information Extraction.* Proceedings of the Lexicon Driven Information Extraction. Frascati, Italy.

Kim J-D., Ohta T., Pyysalo S., Kano Y. and Tsujii J. 2009. *Overview of BioNLP'09 Shared Task on Event Extraction*. NAACL-HLT BioNLP-workshop. Boulder, Colorado.

Kim J-D., Ohta T. and Tsujii J. 2008. Corpus Annotation for Mining Biomedical Events from Literature. *BMC Bioinformatics*, 8;9:10.

Kokkinakis D. (2004). *Reducing the Effect of Name Explosion*. Beyond Named Entity Recognition, Semantic labelling for NLP tasks. Workshop at the 4th LREC. Lissabon, Portugal.

Kokkinakis D. (2009). Lexical granularity for automatic indexing and means to achieve it – the case of Swedish MeSH®. *Information Retrieval in Biomedicine: NLP for Knowledge Integration*. Prince V. & Roche M. (eds). pp. 11-37. IGI Global.

Appendix

Cure

ram	**Cure**
domän	Med
semantisk typ	Cause_change_of_state
kärnelement	**Affliction**, **Body_part**, **Healer**, **Medication**, **Patient**, **Treatment**
periferielement	**Degree**, **Duration**, **Manner**, **Motivation**, **Place**, **Purpose**, **Time**, **Type**
exempel	**Salvan lindrar** även **besvär** som **skavsår**, **sprickor på fingertopparna**, **stickor i fingrarna** samt **skärsår**. **Man** kan **behandla cancer med flera olika metoder**. **Läkaren opererade höger öga** i stället för vänster. **ST-läkaren behandlade henne med höga doser kortison**. **Salvan läker skrubbsår och brännsår**. **Genterapi botade dödssjuka i cancer**. **Transplantation** kan **ha botat hiv-smittad**. **Ljusterapi lindrar** och förebygger **nedstämdhet**, ökar din energinivå och stärker ditt inre lugn.
sms	Type+Treatment, Body_part+Treatment, Medication+Treatment
sms-exempel	Type+LU_EX_ljus.behandling, röntgen.behandling, strål.behandling, värme.behandling Body_part+LU_EX_hjärn.operation, hjärt.operation Medication+LU_EX_kortison.behandling
saldo	**vb:** avvänja..1 behandla..2 bota..1 hela..1 kurera..1 lindra..1 läka..1 läka..2 medicinera..1 operera..1 rehabilitera..1 vårda..1 återanpassa..1 **nn:** behandling..2 dialys..1 diatermi..1 hjärnoperation..1 hjärtoperation..1 huskur..1 läkning..1 kneipkur..1 kortvågsbehandling..1 kur..1 lindring..1 ljusbehandling..1 lobotomi..1 medicinering..1 operation..1 radikaloperation..1 rutinoperation..1 rehabilitering..1 resektion..1 röntgenbehandling..1 skrapning..1 stomi..1 strålbehandling..1 värmebehandling..1 värmeterapi..1 återanpassning..1
saldo (nya)	
kommentar	Obs. behandling förekommer här som både Treatment och det rambärande lexemet, LU.; En sammanflätning (conflation) kan förekomma i objektpositionen mellan Patient och Affliction med vissa verb i den här semantiska gruppen, tex bota hiv-smittad/aids. (Se The Book, s.25 cure epileptic/epilepcy); (I Eng Cure-ramen tolkas 'affliction' som the injuries, disease, pain.);
skapad av	MTG
skapad	2010-01-30
modifierad	2010-02-27

Measuring Risk and Information Preservation: Toward New Metrics for De-identification of Clinical Texts

Lynette Hirschman & John Aberdeen
The MITRE Corporation
202 Burlington Rd.
Bedford, MA 01730
{lynette,aberdeen}@mitre.org

Abstract

Current metrics for de-identification are based on information extraction metrics, and do not address the real-world questions "how good are current systems", and "how good do they need to be". Metrics are needed that quantify both the risk of re-identification and information preservation. We review the challenges in de-identifying clinical texts and the current metrics for assessing clinical de-identification systems. We then introduce three areas to explore that can lead to metrics that quantify re-identification risk and information preservation.

1 Introduction

Our current metrics do not address the questions "how good are current free-text de-identification systems", and "how good do they need to be?" We need measures that quantify *risk of re-identification* based on type and amount of personal health identifier (PHI) leakage (PHI terms not redacted in the de-identification process), and measures that quantify information preservation or readability.

The metrics in current use were developed originally for *entity extraction* (the correct labeling of types of phrases in free text, such as person name, date, organization). Entity extraction performance is typically measured in terms of precision, recall and balanced f-measure at both the token (word) and phrase level. The top de-identification systems (Szarvas, Farkas, & Busa-Fekete, 2007; Wellner, et al., 2007) performed well on these measures, as reported at the first i2b2 De-identification challenge evaluation (Uzuner, Luo, & Szolovits, 2007), achieving accuracies of over 97% token-level f-measure, with recall (sensitivity) of over 95%. Over the past several years, these results have been extended to more record types and record formats; for example, (Friedlin & McDonald, 2008) reported that their MeDS system successfully removed 99.5% of HIPAA-specified identifiers from HL7 records. These are encouraging numbers, but recall and precision do not tell us how good a de-identification system needs to be for a particular intended use.

2 Challenges in Clinical Text

Removing PHI from unstructured text poses new challenges: in contrast to structured information (e.g., fields in a table or a database), we do not know in advance where PHI will appear in a free text record, and we do not know what kinds of PHI will occur. This problem is made more challenging for medical records because the types of record vary greatly in content and in amount of PHI – for example, a lab report will likely contain very little PHI, while a social work note will be likely to contain much more.

Medical records have internal structure that is dependent on the medical record system and the medical record type; there is typically a mix of structured fields (e.g., for patient identifier, patient name, doctor name), along with unstructured fields for free text. This means that de-identification of records must handle a combination of structured and unstructured information. The excerpt below shows two free text fields (CLINICAL HISTORY and IMPRESSION) from a (fictitious) radiology

Proceedings of the NAACL HLT 2010 Second Louhi Workshop on Text and Data Mining of Health Documents, pages 72–75,
Los Angeles, California, June 2010.

report, with several types of PHI, including dates, locations and ages (shown in **bold**).

RADIOLOGY REPORT

CLINICAL HISTORY: Patient is a **4-year 5-month** old male who presented to **Oak Valley Health Center** on **10/11/2007** with a cough of 10 days duration and fever. Patient lives in a densely populated section of **Knoxville**. Rule out pneumonia.

IMPRESSION: Scattered lung densities likely to represent either scattered atelectasis or acute viral illness with no definite lobar pneumonia identified.

This example illustrates how PHI is distributed in the free text portions of a medical record. Removing PHI from these free text portions requires application of techniques from natural language processing that are capable of identifying phrases of specific types based on the lexical content (the words that make up the phrases) and the surrounding words.

3 Current Methods and Metrics

Fortunately, the problem of identifying types of information in free text is a well-studied problem in the natural language processing community. We can leverage several decades of research on *information extraction* and the *named entity identification* problem in particular, including multiple community evaluations such as the Message Understanding Conferences (MUC) (Grishman & Sundheim, 1996) and the subsequent Automated Content Extraction (ACE) evaluations[1] – both focused on extraction from newswire -- as well as evaluations of biomedical entity extraction from the published literature e.g., in the BioCreative evaluations (Krallinger, et al., 2008). In addition, starting in 2006, there have been a series of evaluations for clinical natural language processing, with data sets of clinical records provided by the i2b2 consortium (Uzuner, et al., 2007). It has been critically important to have corpora of medical records, because medical records represent a very different style of text compared to news articles or journal articles. Medical records are characterized by their formulaic and telegraphic style, that is, the use of phrases or incomplete sentences rather than fluent prose, along with heavy use of abbreviations and domain-specific terminology (e.g., "93 yo w NVD"). The systems developed for newswire or for journal articles must be explicitly adapted (or *trained*) to handle the categories required for de-identification as well as the telegraphic language of medical records.

De-identification of free text medical records consists of two steps: recognition and redaction. The phrase recognition stage corresponds to the *named entity recognition* problem mentioned above, namely the ability to identify a sequence of words in running text that constitutes the mention of an entity of a specified type – such as the phrase **Oak Valley Health Center** in the example above. For newswire, types of named entities include person, organization, location, time, date, and money; for biomedical tasks, entities have included genes, proteins, drugs, diseases, etc. For de-identification, the critical elements are the 18 types of protected health information identified by HIPAA,[2] including names, dates, locations, zip codes, phone numbers, social security numbers, ages ninety and above, URLs and other identifying information. Interestingly, most institutions have developed their own set of protected classes of information, e.g., some institutions distinguish between DOCTOR and PATIENT identifiers, which both fall into the more general HIPAA category of NAME.

The techniques used to recognize named entities in text include:

- Lexically-based approaches that rely on matching words (or phrases) against the words or phrases contained in a lexicon;
- Pattern based approaches that are particularly useful for HIPAA-relevant PHI such as telephone numbers, social security numbers, dates, etc.
- Machine learning approaches that are based on statistical models of word sequences. These approaches require *training exemplars* that are used to associate sequences of words with probabilities of types of phrase, e.g., the word(s) following "Dr." or "DR" will likely be a doctor's name.

[1] http://www.itl.nist.gov/iad/mig/tests/ace

[2] Health Insurance Portability and Accountability Act of 1996, Pub. L. No. 104-191, 110 Stat. 1936 (1996).

All three approaches have been used and often combined (Beckwith, Mahaadevan, Balis, & Kuo, 2006; Berman, 2003; Friedlin & McDonald, 2008; Gupta, Saul, & Gilbertson, 2004; Morrison, Li, Lai, & Hripcsak, 2009; Szarvas, et al., 2007; Uzuner, Sibanda, Luo, & Szolovits, 2008; Wellner, et al., 2007) to provide high quality recognition of PHI. The 2006 i2b2 challenge evaluation for automatic de-identification of free text clinical records provided an opportunity for groups to benchmark their automated de-identification systems against a carefully prepared gold standard corpus of medical discharge summaries. The top systems performed well with scores of over 0.97 token-level f-measure and recall (sensitivity) of over 0.95 (Uzuner, et al., 2007).

4 Toward New Metrics

The Uzuner et al. (2007) paper concludes with two important (and as yet unanswered) questions (p. 562):

1. Does success on this challenge problem extrapolate to similar performance on other, untested data sets?

2. Can health policy makers rely on this level of performance to permit automated or semi-automated disclosure of health data for research purposes without undue risk to patients?

We have been particularly concerned with the second question, because it will be very difficult to release automatically de-identified data until we can provide an answer. The metrics used to date have been measures of technology performance, but they do not address the key issues of risk of PHI exposure and readability/preservation of information in the de-identified record.

Recall errors are clearly correlated with risk of PHI exposure, but not all recall errors lead to PHI exposure. For example, the name "Washington, George" might be mistakenly redacted to "LOCATION, NAME" leading to both a recall and a precision error for the word "Washington" but no PHI exposure. Also, some kinds of PHI exposure errors contain much more information (e.g., a patient's last name) than others (a first name; or a telephone extension where the telephone number has been redacted). Friedlin and McDonald (2008) report that MeDS did not miss any full patient identifiers, but it did miss an average of 2.13 patient identifier fragments per report. However, they concluded that none of these fragments were true patient identifiers.

Similarly, precision errors cause mislabeling of results and are correlated with loss of readability. In the extreme case, a system that redacts all words would achieve perfect recall, but very low precision and no information content. A system that replaces real PHI with synthetic (fictitious) PHI might be more resistant to re-identification because it would be difficult for an attacker to distinguish real from fictitious information.

We need new measures that quantify risk of re-identification based on type and amount of PHI leakage (PHI terms or parts of terms not redacted in the de-identification process); and we need measures that quantify information preservation or readability.

We plan to explore three areas that may yield more useful metrics for de-identification. The first is to quantify the re-identification risk through a detailed analysis of PHI in different record types. Given a set of records and a de-identification system, we can generate quantitative data on PHI distribution in different record types, rate of exposure of different classes of PHI (e.g., names vs. locations vs. phone numbers), and likelihood of combinations of exposed PHI. We can distinguish between partial exposure of PHI (e.g., just a first name or just a room number), and combinations of such exposures within a single record (room number and institution provides much more identifying information than room number alone). Using this information, we can develop analyses of risk using methods developed for structured data (Machanavajjhala, Kifer, Gehrke, & Venkitasubramaniam, 2007; Malin, 2007; Sweeney, 2002) by combining statistics from de-identified records with publicly available information (census data, voter registration, etc).

The second area to explore is how to measure information preservation or readability. One approach would be to apply one or more available medical information extraction systems such as the Mayo Clinic cTAKES system (Savova, Kipper-Schuler, Buntrock, & Chute, 2008) to compare information correctly extracted from de-identified data vs. original data. This would provide a reasonable proxy for measuring information loss due to de-identification. Alternatively, Friedlin and McDonald (2008) developed a measure of *interpretability* in their de-identification experiments, defined as preserving test type and test results (for lab reports) or type of report, specimen and conclusion (for pathology reports).

A third area to explore is protection by *hiding in plain sight*. We can determine the reduction in risk from applying resynthesis (Yeniterzi, et al., 2010) to de-identified data, which would have the effect of hiding exposed PHI in plain sight – since such elements would be interspersed with fictitious but realistic looking identifiers (particularly names) inserted as replacements of PHI.

5 Conclusion

Current metrics for de-identification have their origins in information extraction; they neither adequately assess the risk of re-identification, nor do they provide a good measure of information preservation. We plan to address these shortcomings by 1) applying risk analysis methods derived for structured data, 2) using medical extraction systems to assess information preservation, and 3) exploring *hiding in plain sight* protection by using resynthesis to replace identifiers with false by realistic fillers. Once we have alternative measures for risk of re-identification and information preservation, we can also explore the correlation of precision and recall to these new measures. Accurately quantifying and balancing risk of re-identification and information preservation will enable health policy makers to make better decisions about the use of automated de-identification, and sharing of clinical data for research.

References

Beckwith, B. A., Mahaadevan, R., Balis, U. J., & Kuo, F. (2006). Development and evaluation of an open source software tool for deidentification of pathology reports. *BMC Med Inform Decis Mak, 6*, 12.

Berman, J. (2003). Concept-Match Medical Data Scrubbing. *Arch Pathol Lab Med, 127*, 680-686.

Friedlin, F. J., & McDonald, C. J. (2008). A software tool for removing patient identifying information from clinical documents. *J Am Med Inform Assoc, 15*(5), 601-610.

Grishman, R., & Sundheim, B. (1996). *Message Understanding Conference - 6: A Brief History*. Paper presented at the 16th International Conference on Computational Linguistics, Copenhagen.

Gupta, D., Saul, M., & Gilbertson, J. (2004). Evaluation of a deidentification (DE-ID) software engine to share pathology reports and clinical documents for research. *Am J Clin Pathol, 121*(2), 176-186.

Krallinger, M., Morgan, A., Smith, L., Leitner, F., Tanabe, L., Wilbur, J., Hirschman, L., & Valencia, A. (2008). Evaluation of text-mining systems for biology: overview of the Second BioCreative community challenge. *Genome Biol, 9 Suppl 2*, S1.

Machanavajjhala, A., Kifer, D., Gehrke, J., & Venkitasubramaniam, M. (2007). l-diversity: Privacy beyond k-anonymity. *ACM Transactions on Knowledge Discovery from Data (TKDD), 1*(1), 3.

Malin, B. (2007). A computational model to protect patient data from location-based re-identification. *Artif Intell Med, 40*(3), 223-239.

Morrison, F. P., Li, L., Lai, A. M., & Hripcsak, G. (2009). Repurposing the clinical record: can an existing natural language processing system de-identify clinical notes? *J Am Med Inform Assoc, 16*(1), 37-39.

Savova, G., Kipper-Schuler, K., Buntrock, J., & Chute, C. (2008). UIMA-based Clinical Information Extraction System. *Towards Enhanced Interoperability for Large HLT Systems: UIMA for NLP*, 39.

Sweeney, L. (2002). k-anonymity: A model for protecting privacy. *International Journal of Uncertainty Fuzziness and Knowledge Based Systems, 10*(5), 557-570.

Szarvas, G., Farkas, R., & Busa-Fekete, R. (2007). State-of-the-art anonymization of medical records using an iterative machine learning framework. *J Am Med Inform Assoc, 14*(5), 574-580.

Uzuner, Ö., Luo, Y., & Szolovits, P. (2007). Evaluating the state-of-the-art in automatic de-identification. *J Am Med Inform Assoc, 14*(5), 550-563.

Uzuner, Ö., Sibanda, T. C., Luo, Y., & Szolovits, P. (2008). A de-identifier for medical discharge summaries. *Artif Intell Med, 42*(1), 13-35.

Wellner, B., Huyck, M., Mardis, S., Aberdeen, J., Morgan, A., Peshkin, L., Yeh, A., Hitzeman, J., & Hirschman, L. (2007). Rapidly retargetable approaches to de-identification in medical records. *J Am Med Inform Assoc, 14*(5), 564-573.

Yeniterzi, R., Aberdeen, J., Bayer, S., Wellner, B., Clark, C., Hirschman, L., & Malin, B. (2010). Effects of Personal Identifier Resynthesis on Clinical Text De-identification. *J Am Med Inform Assoc, 17*(2), 159-168.

A Comparison of Several Key Information Visualization Systems for Secondary Use of Electronic Health Record Content

*Francisco S. Roque[1], Laura Slaughter[2,3], Alexandr Tkatšenko[4,5]

[1]Center for Biological Sequence Analysis, The Technical University of Denmark, Lyngby, Denmark
[2]The Interventional Center, Oslo University Hospital, Oslo, Norway
[3]Department of Computer and Information Science, Norwegian University of Science and Technology (NTNU), Trondheim, Norway
[4]Institute of Computer Science, University of Tartu, Tartu, Estonia
[5]Software Technology and Applications Competence Center, Tartu, Estonia
http://dsv.su.se/hexanord
*All three authors contributed equally to this work.

Abstract

An overview is provided of six information visualization systems designed specifically for gaining an overview of electronic health records (EHR). The systems discussed all make use of timelines: Lifelines, Lifelines2, KNAVE II, CLEF Visual Navigator, Timeline, and AsbruView. With the exception of Lifelines2, the main user groups targeted are physicians involved in direct patient care. Little attention has been paid towards supporting true secondary use of EHR contents, for activities such as assessing quality of care, patient health and safety monitoring, and clinical trial recruitment. Future work on such systems needs to address the complexity of EHR data, missing and incomplete information, and difficulties in displaying data with differing levels of granularity.

1 Introduction

This paper provides an overview of several information visualization (infovis) systems that have been built for exploring abstracted information from Electronic Health Records (EHR). EHRs are systems that are used to document care of patients. The records can include a wide range of data and information, including medications prescribed and administered, immunization history, laboratory test results, allergies, radiology images, treatment plans, and care notes. Currently, most EHR systems implemented are proprietary and highly customized when used by larger care institutions.

It is usually the case that only clinicians and other healthcare professionals with direct responsibility for providing care have access to patient data. The suggestion of secondary use of health data is not new and has been handled separately from the issue of creating user interfaces and visualizations. Safran et al. discuss the purpose of clinical data repositories in their white paper and point towards the goal of a national framework for the secondary use of health data in the U.S. (Safran et al., 2007). According to their definition, secondary use includes activities such as analysis, research, quality and safety measurement, public health, payment, provider certification and accreditation, marketing, and general business applications, while at the same time taking into account the ethical, political, technical and social implications of such re-use. De Lusignan and van Weel highlight the challenges of making use of clinical data for research, stating, "The available research methods for working with large data sets are limited; it is difficult to infer meaning from data; there is a rapid pace of change in both medicine and technology; and integrating data without reliable unique identifiers is difficult." (de Lusignan and van Weel, 2006). Prokosch and Ganslandt have recently summarized the latest advances in enabling clinical data re-use for research purposes (Prokosch and Ganslandt, 2009). They identify as key challenges the establishment of comprehensive clinical data repositories, the establishment of professional IT infrastructure to sup-

Proceedings of the NAACL HLT 2010 Second Louhi Workshop on Text and Data Mining of Health Documents, pages 76–83,
Los Angeles, California, June 2010.

port clinical data capture, and the integration of medical record systems and clinical trial databases. As discussed in these articles, aggregated, abstracted and manipulable information is underutilized and hard to come by.

The emerging field of *Visual Analytics* (Keim, 2008) is relevant to this review. This field is focusing on combining related research areas such as visualization, data mining and statistics to handle large and heterogeneous volumes of data, such as EHR. The systems we encountered are integrating human judgment with automated analysis, suggesting that future work will be related to handling massive amounts of data that contains missing elements - including the results of textual analysis of records content.

1.1 Purpose

Our motivation for creating this overview is to compare and discuss some of the available information visualization/visual analytics tools and how are these used for secondary, i.e. for purposes other than direct patient care. This is a first step towards infrastructure and coordinating efforts to produce systems that are based on standard input formats, and meet the needs of specifically defined users. The reader of this overview is most likely working on information extraction, temporal abstraction, and summarizing EHRs.

Source	*Search Keywords*
Pubmed	visualization health records Medical Records Systems, Computerized Computer Graphics User-Computer Interface
ACM DL	electronic health records or medical record information visualization or visualization healthcare or health care user interface
IEEE DL	visualization medical records
Google Scholar	electronic medical records or EHR information visualization visual analytics

Table 1. Keywords searched.

1.2 Scope

The review is non-systematic. We didn't expect to find large numbers of articles, since this is a relatively narrow area of interest. The search was confined to user interfaces and visualizations for EHR data, we searched pubmed, ACM digital library, IEEE library, and Google Scholar, using basic keywords and checked references in found articles. We also looked for papers on work we had read or known about previously from conferences or other sources. The literature search covered articles in English only. Keywords used are listed in Table 1.

2 Systems

In this section, we give an overview of the state-of-the-art systems related to visualization of temporal information in EHRs. Our intention is to cover broad areas of application including representation of medical histories, visual data query and aggregation, generation of temporal abstractions and visualization of treatment plans. Due to the limitations in space, we focus only on the most representative systems, which feature interesting and potentially reusable visualization techniques.

Lifelines

LifeLines uses a timeline visualization technique to represent personal histories, medical records and other types on biographical data (Plaisant et al., 1996). In LifeLines, horizontal bars are used to depict temporal duration and location of events on a horizontal time axis. Similar events are organized into facets, which can be expanded and collapsed to provide increasing or decreasing level of detail. Color notations and line thickness are used to indicate the importance and relationship of events. To handle regions with high data density, LifeLines provides zooming functionality allowing users to compress and stretch the time scale at any location. Additional content (e.g., multimedia) can be added in a linked fashion. Authors apply LifeLines in the analysis of complex patient medical records to visualize temporal relationships between treatments, consultations, disorders, prescriptions, hospitalizations and other events.

Lifelines2

LifeLines2 (Wang et al., 2008) is an extension of LifeLines, allowing the user to analyze records from multiple patients at a time. The system facilitates comparative visualization of records by means of aligning, filtering and sorting operations. By aligning patient records on some common reference event (e.g., the first heart attack), users can easily spot co-occurring and neighboring events. Ranking and filtering operations complement alignment by interactively reordering or narrowing the set of records to suit a user's changing focus. The system proved to be particularly suitable for observational research, where researchers analyze data from different studies in order to better understand health problems or study the effect of treatments, and in finding patients for clinical trials. Evaluation studies showed that the system significantly simplifies typical analytical tasks and that medical specialists can quickly learn the interface. LifeLines2 is currently used to display EHR data provided by the Informatics for Integrating Biology & the Bedside (i2b2) Project (Murphy et al., 2006).

While in LifeLines2 the main focus is on visualizing temporal ordering of events, Wang et al. emphasizes practical need in viewing multiple records as an aggregate in order to study frequency of event data over time (Wang et al., 2009). For instance, a user might be interested to analyze blood pressure of all patients who have had an open-heart surgery within 3 months of their first heart attack. As a solution, authors complement LifeLines2 framework with a new visualization technique, called temporal summaries, which represents distributional trends of events over a set of records in a histogram-like chart. Furthermore, the system allows splitting the whole dataset of records into multiple subsets and use temporal summaries to compare event patterns between these groups.

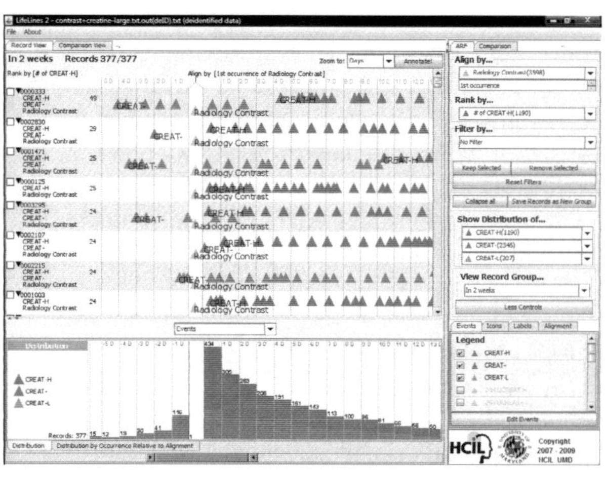

Figure 1. The Lifelines2 main window, with focus on timelines.

CLEF

Hallet (Hallett, 2008) proposes a visualization architecture for browsing medical histories, which integrates visual navigation tools and automatically generated textual summaries. While the graphical interface facilitates interactive navigation, textual descriptions can, in addition, convey complex temporal information and display details that would otherwise be too complex for visualization components. Within the system, the patient's medical history is represented as a network of semantically and temporally organized events, which serves as an input for visualization and natural language generation components. The visual navigator depicts a high level overview of a patient's medical history by plotting events along three parallel timelines, corresponding to diagnoses, treatments and investigations. In addition to zooming time scale and detail-on-demand functionality, the navigator provides interactive visualization of semantical relationships between events (e.g., caused-by, has-locus, indicated-by, etc.). Having different features from the LifeLines interface, the navigator also allows the user to visualize numerical data (e.g., results of blood tests) by plotting results of measurements on separate line charts. Natural language generation is used for two purposes: 1) to create customized textual reports for printing or exchange purposes and 2) as a support tool for the visual navigator, to enable better description of complex events and relationships between them.

KNAVE-II

KNAVE-II (Goren-Bar et al., 2004) is an interface enabling knowledge-based visualization and interactive exploration of time-oriented data at different levels of temporal abstractions (e.g., abstraction of periods of bone marrow toxicity from raw individual hematological data). Users can navigate through the links of a semantic network while simultaneously navigating visually through multiple degrees of temporal abstraction of the dataset under observation. The evaluation results have shown that users of KNAVE-II were able to perform queries both faster and more accurately than with other standard tools.

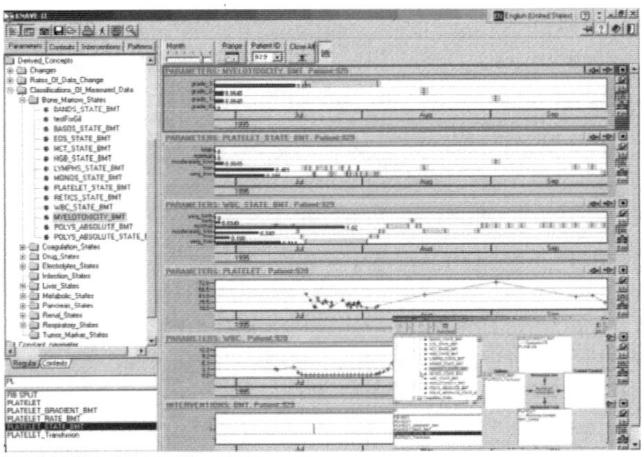

Figure 2. The Knave-II system.

TimeLine

The TimeLine system (Bui et al., 2007) is a problem-centric temporal visualization of patient records. The contents of the EHR are integrated, reorganized, and displayed within the user interface (UI) along a timeline. It is similar to Lifelines in the way that the different elements of the EHR are grouped along the y-axis: imaging, reports, lab tests, etc are collapsible categories. However, unlike Lifelines, the TimeLine system uses an XML data representation to handle data from distributed, heterogeneous medical databases. Data elements that are displayed in the UI are classified based on a knowledge base that guides both data inclusion rules and the visualization metaphors used to render the data.

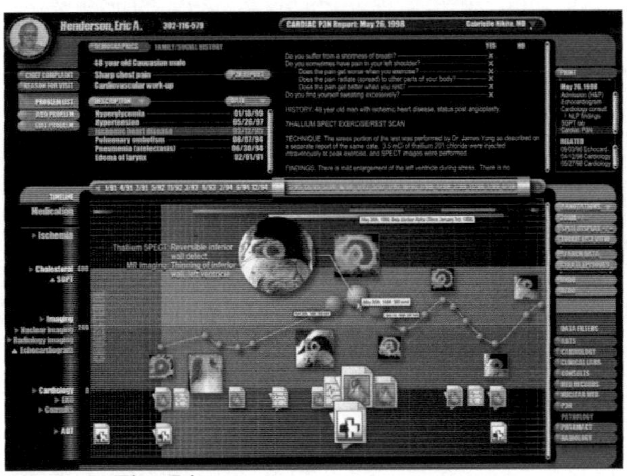

Figure 3. TimeLine system.

ASBRUVIEW

AsbruView (Kosara and Miksch, 2001) is a visualization and user interface on top of Asbru language (Shahar et al., 1996) designed to represent treatment procedures as structured time-oriented plans. AsbruView represents hierarchical and temporal relationships between treatment plans using a 3D visualization perspective. Plans are aligned along the time axis and can be stacked on top of each other and laid out in different ways. To simplify the interface, all graphic elements are represented by well-known real world objects (e.g., track, traffic light, etc.). Also a 2D view is available which focuses on temporal aspects of plans in greater detail. To depict uncertainty of future events, AsbruView extends the timeline by using time annotation glyphs (Chuah, 1997).

3 Comparisons

Infovis techniques are a way of augmenting human cognitive capabilities, to help humans find patterns in large volumes of data. The systems described above target specific user types that will benefit from the visualization methods. While some user interfaces were developed in close dialog with medical practitioners, like Lifelines2 and Knave-II, others, such as the first Lifelines, Clef and Asbruview have had only minimal input from their intended audience.

3.1 Users, Goals and Tasks

Most of the tools were directed at clinicians and clinical practice, although they were not always developed in close relation to them. Table 2 gives an overview of intended users for each of the named systems, and their proposed goals/tasks. From the user point of view, a number of tasks and goals can be defined for each tool. Some are very specific and tend to care for niche usages, while others provide more general visualization methods that can be applied to a number of situations.

System	*Users, Goals, Tasks*
Lifelines	Clinician Patient care Use EHR content in temporal time-based view
Lifelines2	Clinical researchers Research Compare patterns of events, detecting trends
CLEF	Clinician, Biomedical researcher Patient care Visualize timelines, use NLP to extract complex temporal data, aggregate numerical data
KNAVE-II	Clinician

	Patient care Generation and exploration of context sensitive abstractions of temporal data
TimeLine	Clinician Patient care Use EHR content in temporal time-based view, with additional filters on data based on NLP techniques
AsbruView	Clinician Patient care Medical therapy planning and execution

Table 2. Users, Goals, Tasks.

These systems were designed with input from only a few medical personnel involved in the project. In general, articles we read concerning these systems that have a more guided development process, i.e. closely related with physicians, have more specific goals and tasks, because they were designed with these in mind. Visualizing data for decision-making and analyzing treatment outcome is often a general goal in many of the tools developed in interaction with medical staff (Aigner and Miksch, 2006; Mamykina et al., 2004; Portet, 2009). There is an emphasis on pre-processed patient data, specifically numeric, such as lab tests, heart rate, and blood pressure. Systems mainly try to help physicians answer questions about correlations in the patient's data, and provide a means for supporting quick decision-making making when combining several types of highly heterogeneous data. Physicians can follow a specific treatment plan and check the patient's physiological variables over time. This also enables the practitioner to check the influence of certain variables in the treatment process and change the protocol if needed. CLEF, for example, allows the physician to discover events during specific time spans, such as searching for past specific liver problems. Lifelines2 is specifically geared towards research uses and towards answering complex queries. In Lifelines2, a case study involved verifying the results of a clinical study with real-life EHR data to see if clinical care data differ from the study results.

The systems we discuss have conducted evaluation studies as a part of the end-stages of development. The Lifelines evaluations were conducted in another domain (use of pattern searching related to monitoring graduate student progress), with limited interviews and input from experts in the medical domain. The KNAVE system conducted a cross-over study with doctors, comparing KVAVE with existing tools. TimeLine was evaluated following the development of the interface by five radiologists- focusing on questions related to data integration and the temporal display. Asbruview was evaluated using questionnaires sent to clinicians.

3.2 Visualization Methods

The focus of this paper is on temporal visualization methods since this has been the primary visualization type studied for aiding humans in organizing and exploring patterns in abstracted EHR content. All the systems that are compared in this paper display some type of timeline with time running from the left part the screen to the right, time being on the x-axis, and categories of events along the y-axis. Various techniques for graphically representing specific events are used (e.g. icons, shapes), AsbruView makes use of 3D, while all the others are flat 2D.

Infovis has been the keyword used to describe these systems, with the idea of presenting a method for human users (most often stated as being clinicians), to recognize patterns and thereby "amplify cognition" (Chittaro, 2006). Other methods for recognizing patterns in EHR for secondary use are purely automated and conducted through data mining techniques. Bertini and Lalanne (2009) wrote about the complementary role of automatic data analysis and visualization in knowledge discovery. They discuss "visual analytics", an outgrowth of infovis that can be seen as an integrated approach combining visualization, human factors, and data analysis. They suggest 4 categories for classifying approaches: Pure Visualization (VIS), Computationally-enhanced Visualization (V++), Visually enhanced Mining (M++), and Integrated Visualization and Mining (VM). In the systems we have compared, there is a spectrum of ideas about how to visualize EHR contents, including movements towards "enhanced" or "intelligence" in the processing of the underlying EHR data. In Lifelines2, the data visualized was obtained from anonymized EHRs though cooperation with the i2b2 Project (Murphy et al., 2006) The input form of the data is a simple 3-column table containing "ID", "Event Type", and "Time". Each ID can have multiple events happening at various times. Lifelines2 allows sorting of the data so that records with the most incidents of one type of event are shown at the top of the screen. This type of infovis relies on human pattern recognition only and would be considered as "VIS" by Bertini and Lalanne (2009). In the CLEF project, the CLEF Chronicle, which underlies the visualizations, is a semantic network modeling of what happened to the patient, why, and how. Semantic relations are: causality, reason, finding, and consequence. The types of events modeled are: problem, investigation, and treatment. The CLEF Visual Navigator might be considered as "V++", computationally enhanced visualization because some sort of automated computation supports the visualization. In CLEF, the visual display is "enhanced with visual techniques for highlighting relationships between events on the timeline." None of the systems so far that we have seen, would qualify as "visually enhanced mining" or "integrated visualization and mining." Table 3 provides a full overview for all systems reviewed.

System	*Category*	*Notes*
Lifelines	VIS	
Lifelines 2	VIS	
CLEF	V++	• automated generation of summaries • semantic network of EHR record events
KNAVE-II	V++	• semantic (ontology-based) navigation and exploration of the data • knowledge base is used to interpet raw data
TimeLine	V++	• data mapping and reorganization • content-based techniques to elucidate predominant subject of reports for classification
AsbruView	VIS	

Table 3. Visual Analytics of Systems using Bertini and Lalanne's (2009) classification.

The papers we have read that cover EHR visualization, as seen in the systems presented, express the complexity of abstracted EHR data. Missing and inconsistent data, dealing with hierarchical data, and problems with granularity are all concerns that become readily apparent through attempting to build infovis systems. Wang (2008) summed it up best "Clinical data tend to be messy with aspects that become only obvious when the data is visualized. The same heart attack might be recorded three times in three days (by the emergency room physician, a cardiologist, and a clerk from the billing office) and it can be hard to differentiate it from 3 separate events. Even if medical event information is carefully recorded at the time of the doctor visit or during a hospitalization, the time stamp is usually inaccurate by nature." Future work on visualizations needs to adequately address the complexity of the data rather than work with test data that is too simplistic.

3.3 Text Mining Tasks

All mentioned systems, except the CLEF and TimeLine, operate with readily available lists of type- and time-tagged events. However, clinical records are often stored in textual form what makes them inaccessible for machine processing. Text mining techniques need to be applied to automatically transform textual data into structured, normalized form. Key tasks involve event extraction, classification and normalization.

The CLEF system uses an advanced information extraction engine to identify pre-defined classes of entities (e.g. diseases, investigations, problems, drugs, etc.) and semantic relationships between them (e.g. investigation indicates problem) in natural language texts. The information extraction process involves lexical and terminological analysis, syntactic and semantic analysis, and discourse analysis. To address the complexity of medical language, the system makes use of language resources including the Unified Medical Language System and the Gene Ontology. Extracted information is stored in templates, which can be queued or used to generate textual summaries. The TimeLine system makes use of both textual contents of the EHR as well as numerical data and codes. An NLP-based system is used in conjunction with the TimeLine UI, for example, performing section analysis in radiology reports to determine whether specific subsections exist within the reports that are related to certain medical problems (Bui et al. 2007).

4 Conclusions

The infovis systems analyzed allow secondary use of EHR content data especially aimed at clinicians documenting patient care. All of them are focused on visualizing temporal data in a timeline, while displaying specific events from the patient data.

Although directed at medical practitioners in their daily patient care routine, they were not always developed with user feedback. Evaluation of the different tools was often based on situations outside of the clinical setting, and might not reflect reality. A more intimate dialog with clinicians would benefit the creation of targeted systems addressing specific needs of the medical community.

The overall goal of these tools is to present users temporal information contained in a record, improving their ability to recognize patterns for knowledge discovery and following treatment. They introduce simple visualization tools, but some include automated computational enhancements supporting it.

EHR contain missing and inconsistent data, which is in general messy. Due to the complexity of the underlying data, future work needs to address these intricacies rather than using simplistic approaches.

Acknowledgments

We would like to thank Nordforsk and the Nordic Council of Ministers for the funding of our research network HEXAnord - Health text Analysis network in the Nordic and Baltic countries. This work was partially supported by a grant from the Villum Kann Rasmussen fund.

References

Aigner, W., Miksch, S., 2006. CareVis: integrated visualization of computerized protocols and temporal patient data. Artif Intell Med 37, 18.

Bertini, E., Lalanne, D., 2009. Surveying the complementary role of automatic data analysis and visualization in knowledge discovery, Proceedings of the ACM SIGKDD Workshop on Visual Analytics and Knowledge Discovery: Integrating Automated Analysis with Interactive Exploration. ACM, Paris, France, pp. 12-20.

Bui, A., Aberle, D.R., Kangarloo, H., 2007. TimeLine: Visualizing Integrated Patient Records. IEEE Transactions on Information Technology in Biomedicine 11(4): 462-473.

Chittaro, L., 2006. Visualization of patient data at different temporal granularities on mobile devices, AVI '06: Proceedings of the working conference on Advanced visual interfaces, pp. ACM--487.

Chuah, M.a.E., S., 1997. Glyphs for software visualization. 5th International Workshop on Program Comprehension (IWPC '97) Proceedings, 183-191.

de Lusignan, S., van Weel, C., 2006. The use of routinely collected computer data for research in primary care: opportunities and challenges. Fam Pract 23, -263.

Goren-Bar, D., Shahar, Y., Galperin-Aizenberg, M., Boaz, D., Tahan, G., 2004. KNAVE II: the definition and implementation of an intelligent tool for visualization and exploration of time-oriented clinical data, AVI '04: Proceedings of the working conference on Advanced visual interfaces, pp. ACM--174.

Hallett, C., 2008. Multi-modal presentation of medical histories, IUI '08: Proceedings of the 13th international conference on Intelligent user interfaces, pp. ACM--89.

Keim, D.A., Mansmann, F., Schneidewind, J., Thomas, J., Ziegler, H., 2008. Visual analytics: Scope and challenges. Visual Data Mining: Theory, Techniques and Tools for Visual Analytics, Springer, 76-90.

Kosara, R., Miksch, S., 2001. Metaphors of movement: a visualization and user interface for time-oriented, skeletal plans. Artif Intell Med 22, 111-131.

Mamykina, L., Goose, S., Hedqvist, D., Beard, D.V., 2004. CareView: analyzing nursing narratives for temporal trends, CHI '04: CHI '04 extended abstracts on Human factors in computing systems, pp. ACM--1150.

Murphy, S.N., Mendis, M.E., Berkowitz, D.A., Kohane, I., Chueh, H.C., 2006. Integration of clinical and genetic data in the i2b2 architecture. AMIA Annu Symp Proc.

Plaisant, C., Milash, B., Rose, A., Widoff, S., Shneiderman, B., 1996. LifeLines: Visualizing Personal Histories, CHI, pp. -227.

Portet, F., Reiter, E., Gatt, A., Skyes, C., 2009. Automatic generation of textual summaries from neonatal intensive care data. Artificial Intelligence, 789-816.

Prokosch, H.U., Ganslandt, T., 2009. Perspectives for medical informatics. Reusing the electronic medical record for clinical research. Methods Inf Med 48, -44.

Safran, C., Bloomrosen, M., Hammond, W.E., Labkoff, S., Markel-Fox, S., Tang, P.C., Detmer, D.E., Panel, E., 2007. Toward a national framework for the secondary use of health data: an American Medical Informatics Association White Paper. J Am Med Inform Assoc 14, -9.

Shahar, Y., Miksch, S., Johnson, P., 1996. An intention-based language for representing clinical guidelines. Proc AMIA Annu Fall Symp, 592-596.

Wang, T.D., Plaisant, C., Quinn, A.J., Stanchak, R., Murphy, S., Shneiderman, B., 2008. Aligning temporal data by sentinel events: discovering patterns in electronic health records, CHI '08: Proceeding of the twenty-sixth annual SIGCHI conference on Human factors in computing systems, pp. ACM--466.

Wang, T.D., Plaisant, C., Shneiderman, B., Spring, N., Roseman, D., Marchand, G., Mukherjee, V., Smith, M., 2009. Temporal Summaries: Supporting Temporal Categorical Searching, Aggregation and Comparison. IEEE_J_VCG 15, -1056.

Machine learning and features selection for semi-automatic ICD-9-CM encoding

Julia Medori
CENTAL
Université catholique de Louvain
Place Blaise Pascal, 1
1348 Louvain-la-neuve
julia.medori@uclouvain.be

Cédrick Fairon
CENTAL
Université catholique de Louvain
Place Blaise Pascal, 1
1348 Louvain-la-neuve
cedrick.fairon@uclouvain.be

Abstract

This paper describes the architecture of an encoding system which aim is to be implemented as a coding help at the *Cliniques universtaires Saint-Luc*, a hospital in Brussels. This paper focuses on machine learning methods, more specifically, on the appropriate set of attributes to be chosen in order to optimize the results of these methods. A series of four experiments was conducted on a baseline method: Naïve Bayes with varying sets of attributes. These experiments showed that a first step consisting in the extraction of information to be coded (such as diseases, procedures, aggravating factors, etc.) is essential. It also demonstrated the importance of stemming features. Restraining the classes to categories resulted in a recall of 81.1 %.

1 Introduction

This paper describes a series of experiments carried out within the framework of the CAPADIS project.[1] This project is the product of a collaboration between the UCL (Université catholique de Louvain, Belgium) and the Cliniques universitaires Saint-Luc. Saint-Luc is one of the major hospitals in Belgium. Each year, a team of file clerks processes more than 85,000 patient discharge summaries and assigns to each of them classification codes taken from the ICD-9-CM (International Classification of Diseases – Ninth Revision – Clinical modification) (PMIC, 2005).

The encoding of clinical notes (or patient discharge summaries) into nomenclatures such as the International Classification of Diseases (ICD) is a time-consuming, yet necessary task in hospitals. This essential process aims at evaluating the costs and budget in each medical unit. In Belgium, these data are sent to the National Health Department so as to compute part of the hospital's funding.

Our aim is to help coders with their ever-growing workload. More and more patients' stays need to be encoded while the number of coders remains the same. Our goal is therefore to develop an semi-automatic encoding system where the role of the coders would be to check and complete the codes provided by the system.

This paper focuses on machine learning methods as automatic encoding techniques. More specifically, it focuses on the appropriate set of attributes to be chosen in order to optimize the results of these methods.

It will therefore present the structure of the system and compare the results of different inputs to the machine learning approach. Section 2 gives a more detailed description of the objectives of this project. Section 3 gives an overview of the architecture of the system: first, the extraction part will be described, and then, the automatic encoding stage will be discussed. Section 4 will focus on the machine learning experiments that

[1] http://www.iwoib.irisnet.be/PRFB/t10/t10_medori_fr.html

Proceedings of the NAACL HLT 2010 Second Louhi Workshop on Text and Data Mining of Health Documents, pages 84–89,
Los Angeles, California, June 2010. ©2010 Association for Computational Linguistics

were conducted. The results will be presented and discussed in sections 5 and 6.

2 Objectives

Since the early 1990s and the rise of the computational linguistics field, many scientists have looked into the possible automation of the encoding process (Ananiadou and McNaught, 2006; Ceusters et al., 1994; Deville etal., 1996; Friedman et al., 2004; Sager et al., 1995; Zweigenbaum et al., 1995). Two different approaches distinguish themselves from one another: a symbolic approach as in (Pereira et al., 2006) and a statistical one. Both methods scored highly in the "Computational Medicine Challenge" (CMC) organized by the "National Library of Medicine" in 2007 (Pestian et al., 2007): among the best three systems, two combined a statistic and a symbolic approach and only one relies only on a symbolic approach. Most systems participating took a hybrid approach as in (Farkas and Szarvas, 2008).

During ACL 2007, Aronson (2007) presented within the framework of the same challenge, four different approaches, symbolic, statistical and hybrid. His conclusion was that combining different methods and approaches performed better and were more stable than their contributing methods. Pakhomov (2006) describes Autocoder, an automatic encoding system implemented at Mayo Clinic that combines example-based rules and a machine learning module using Naïve Bayes.

Within the scope of this challenge, only a limited number of codes were involved.

The objective of our work is to build such a tool to help the team of coders from the *Cliniques Universitaires Saint-Luc*. Three facts are noteworthy: the clinical notes we work on are written in French; they originate from all medical units; and all the codes from the ICD are used in the process (around 15,000). Most studies are limited on at least one of these criteria: most systems are developed on English as more language resources are available, and they often focus on specific types of notes, e.g. the CMC focused on radiology reports.

3 System description

The system is divided into two units: an extraction unit which aims at marking up information considered as relevant in the encoding process, and an encoding unit which, from extracted information generates a list of codes.

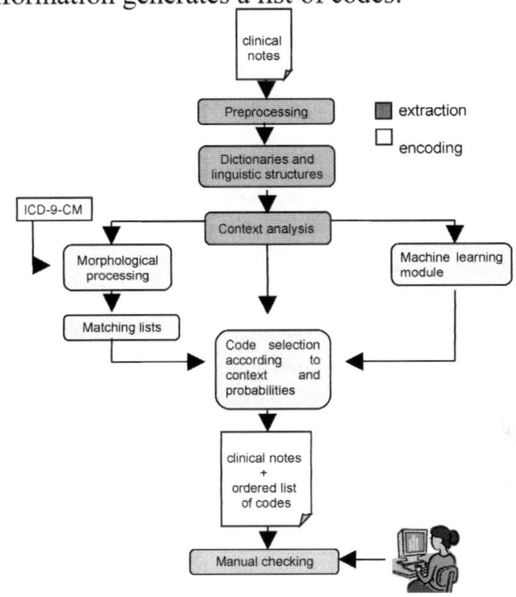

Figure 1. System structure

Extraction: The system aims at reproducing the work of human coders. Coders first read the text, extract all the pieces of information that have to be encoded, and 'translate' information into codes of the ICD-9-CM. The idea behind our tool is to recreate this process. The main source of information coders use are the patient discharge summaries written by doctors summarizing all that happened during the patient's stay: diagnoses, procedures, as well as the aggravating factors, the patient's medical history, etc. These files are electronic documents written in free text with no specific structure.

We developed a tool which aims at extracting the necessary information from these texts: terms referring to diseases but also anatomical terms, the degree of seriousness or probability of a disease, aggravating factors such as smoking, allergies, or other types of information that may influence the choice of a code.

There are many ways of referring to the same diagnosis or procedure, we therefore needed to build specialized dictionaries that would comprise as many of these wordings as possible. The dictionaries of diseases and procedures were mainly built automatically using the UMLS and the classifications in French it comprises. Other specialized dictionaries (anatomical terms, medical departments, medications, etc.) were developed from existing lists. These then were gradually completed manually as the development of the extraction tool went on.

However, the plain detection of terms is not sufficient. It is important to detect in which context these terms occur. For instance, a diagnosis that is negated will not be encoded. The identification of contexts required the use of finite-state automata and transducers. These transducers are represented by graphs that describe the linguistic structures indicating specific contexts. These graphs were hand crafted using the UNITEX software tool[2] (Paumier, 2003). An example of a graph matching fractures and sprains is presented in figure 2.[3] Each path of the graph describes a recognized linguistic structure.

Graphs were also used to broaden the scope of the terms detected by dictionaries. For instance, not only do diseases need to be extracted but, to code, one also needs to know which part of the body is affected.

Certain types of diagnoses also have to be described via graphs such as smoking as there are many ways in which to say that someone smokes or not. Ex: "he smokes 3 cigarettes a day." "He used to smoke." "Occasionally smokes." "Heavy smoker." "Does not smoke."

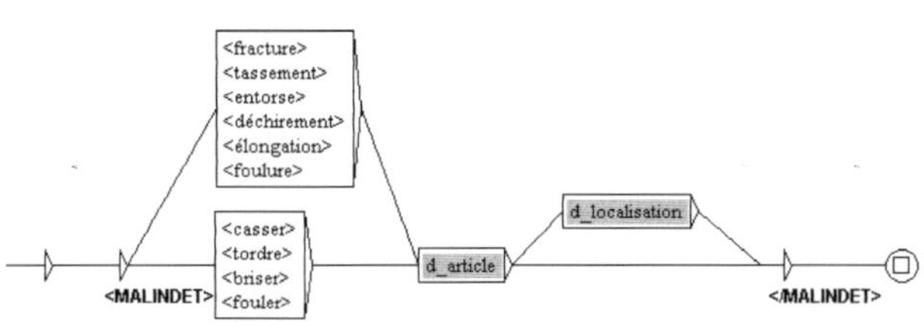

Figure 2. Example of a UNITEX graph matching patterns such as fractures and sprains.

[2] http://www-igm.univ-mlv.fr/~unitex/

[3] The grey boxes indicate calls to other graphs. Here, *d_localisation* is a graph matching anatomical terms.

Our aim was to develop a wide-coverage system. We therefore focused mainly on the General Internal Medicine service in order to develop the grammars and dictionaries. It is a very diverse department where physicians have to face all kinds of diseases.

The graphs and dictionaries on which is based our extraction system were built during the first phase of the project. A more detailed description and evaluation of the extraction process can be found in (Medori, 2008).

Encoding: As was said above, two main approaches to the encoding problem coexist: the symbolic approach and the statistical approach. Both have their benefits and drawbacks. The symbolic approach is a time-consuming approach as it involves describing linguistic rules linking text to diseases. The statistical approach has the advantage of being fast to compute but the need for a large amount of data often hampers the use of these methods. However, both methods give reliable results, and a combination of both is the option generally favored. In our context, we chose to combine both approaches as a large corpus of clinical notes is at our disposal.

Saint-Luc provided us with a corpus of 166,670 clinical notes. The codes that were assigned to them by the coders were also provided. This corpus gives us the chance to develop and test statistical methods in a 'real life' experiment.[4]

However, whatever the results, we will need to combine these methods with linguistic rules. There are two main reasons for this : in the near future, we will have to face the problem of having to switch to another classification. The change to ICD-10-CM is planned for 2015. Therefore, at that time, we will not have enough learning data to be able to generate the list of codes in a statistical manner. The second reason is that there are codes that are seldom assigned and for which we will not have enough occurrences in our corpus to be able to extract them statistically.

This paper focuses on the statistical tests that were conducted on our corpus. An insight into a symbolic method using the matching of morphemes can be found in (Medori, 2008).

[4] In this paper, the experiments were conducted on a smaller corpus. At a later stage, the methods chosen for the final system will need to be trained on the full corpus.

4 Experiment

As a first encoding experiment, we chose to focus on a baseline machine learning method: Naïve Bayes. This method has often been used and proves to be robust.

To conduct this experiment, we used Weka, a data mining software[5] developed at the University of Waikato. For more information on this tool, see (Witten, 2004).

In order to test this method we took a sub-set of 19,994 discharge summaries from the General Internal Medicine department. In order to test how necessary the extraction step is, we chose the texts from the department on which the development of the extraction rules were based.

These notes were assigned 102,855 codes which makes up 4,039 distinct codes.

This corpus was then divided into two subsets: 90% of the 19,994 patient discharge summaries were used as the training corpus and 10% as the test set.

As with any machine learning method, enough data for each class is needed in the training set in order to be able to classify correctly. Therefore, we built a classifier for each code that was manually assigned at least 6 times in our corpus. This resulted in 1,497 classifiers, which means that we did not have enough data to be able to assign 2,542 codes which make up 5% of all the assigned codes.

Four experiments were conducted:

Experiment 1. In our first experiment, the selected attributes were the terms that were highlighted as diagnoses by the extraction step. The diagnoses identified in a negative context were removed from the features list. These resulting list of extracted terms went through a normalization process: accents and stop words were removed; words were decapitalized.

Experiment 2. The second experiment aimed at proving the relevance of the stemming of these terms. The attributes in this experiment were therefore the terms that were extracted, then normalized and stemmed using Snowball Stemmer[6] which is an implementation of the Porter algorithm.

Experiment 3. In this third experiment, we wanted to check the relevance of the extraction process (see experiments 1 and 2). Therefore, the attributes comprised all the words contained in the clinical notes apart from stop words. The words were stemmed in the same way as the extracted terms in experiment 2.

Experiment 4. In all the previous experiments, the classes to be assigned consisted in codes. In this experiment, classes are reduced to categories of codes: represented by the first three digits of a code. The attributes are the same as in experiment 1: extracted terms (non-stemmed). As the system is designed as a coding help i.e. its aim is to generate a list of suggested codes, and not as a fully automated encoding system, one could imagine listing categories of codes instead of codes themselves and then let the coders look up in the hierarchy for the appropriate code within the selected category.

At the end of each experiment, we end up with a list of the 1,497 codes from ICD-9-CM ordered by their Naïve Bayes score for each letter.

The measure that is most interesting here is the recall. The list of suggested codes needs to comprise most of the codes the coder will need so that he/she does not have to go elsewhere to find the appropriate code. Therefore, we kept three measures of recall. It is important to keep the list of codes to be presented to the user short and manageable. Larkey and Croft (1995) used the same measures and set the limit number of codes to 20. This choice is arbitrary but seems like a sensible limit. In Saint-Luc, the maximum number of codes a file clerk can assign to a patient discharge summary is 26 (the principal diagnosis is assigned the letter A and all the other codes are ordered according to the other letters of the alphabet). However, few reports are actually assigned 26 codes (15 out of 19,994). The average number of codes assigned by the file clerks in our set of 19,994 discharge summaries is 6.2.

The three measures of recall are **Recall10**, **Recall15** and **Recall20** which are the measures of micro averaged recall if we show the first 10, 15 and 20 most likely codes respectively.[7]

---

[5] http://www.cs.waikato.ac.nz/~ml/weka/

[6] http://snowball.tartarus.org/

[7] It should be noted that we keep in the list of suggested codes all the codes that tie last with the 10th, 15th and 20th position respectively.

5 Results

The results of the experiments described above are detailed in figure 3.

	Rec10	Rec15	Rec20
1(att: extracted terms)	50.4	56.4	60.5
2 (att: stemmed extracted terms)	56.1	64.1	69.1
3 (att: all words, stemmed)	39.1	40.3	41.4
4 (att: extracted terms classes : categories)	64.0	75.1	81.1

Figure 3. Recall for each experiment (in %)

Experiment 1. From the results of this baseline experiment, considering the extracted terms and retaining the 20 most likely codes according to the Naïve Bayesian statistics, more than 60% of the codes manually assigned to the test notes can be found in this list.

Experiment 2. The stemming of the extracted terms increased the recall by 8.6%.

Experiment 3. If considering all the words as attributes, the recall when retaining 20 possible candidates is around 40% while when attributes are selected through the extraction process, the recall increases to 69% which is an increase of about 28%. This result proves that the extraction process is an essential step in the system and clearly improves the performance of the statistical encoding unit.

Experiment 4. When classes are limited to categories, Recall20 jumps to 81.1% which is 20.6% more than in experiment 1 which was conducted with the same attributes but where classes were codes. This supports our idea that showing a list of categories instead of codes could be an interesting alternative for coders: they would be shown more codes while keeping the list manageable, and then could browse easily into the sub-structure of the classification.

6 Discussion

The choice of attributes is important when testing machine learning methods. In the framework of the development of an encoding system, we proved that a first step consisting in selecting the terms carrying the information that needs to be encoded is essential. We also showed that the use of a simple stemming algorithm clearly improves the performance of the method.

In the last experiment, classifying the clinical notes by categories of codes resulted in a recall of 81.1%. This reinforces our opinion that, to make sure that all the needed codes are present for the coder, we could list categories and let him/her browse through the codes from there.

It is important to put these results in light of where the codes originate. Most of the information that needs to be encoded is present in these clinical notes. However, even though efforts are made in order for this to change, many physicians still do not compile all the information into these notes. Coders therefore still have to look up into the whole patient record in order to find additional codes. The proportion of codes that cannot be inferred from the clinical notes can be very high. A study conducted by Sabine Regout, a patient discharge summary specialist in Saint-Luc, on 250 clinical notes from 25 medical units, showed that in most departments, 15 to 20% of the codes assigned by the clerks cannot be inferred from the notes. This proportion can increase up to 80% in some surgery departments. This evaluation proves that without a change of mind-set from the physicians, our system can only aim to be a coding help for file clerks. Analyzing all the different types of documents contained in patient records would be a difficult task as they comprise a variety of documents with different structures and formats, and some of them are hand-written documents. For our experiments, this also means that the maximal recall value we will be able to get is around 80%.

In these experiments, we were not able to check the inter-annotator agreement but we must keep in mind that, as in any classification task where humans set the gold-standard, one must expect some degree of errors and variation in the coding.

Another observation influences the maximal number of codes we will be able to retrieve is that we built classifiers for all the codes for which we had enough data. This lead to the building of 1497 classifiers. This represents 95% of all the codes assigned to our test notes. This decreases our maximal recall value by 5%.

The codes that are seldom assigned will therefore never show up in our list of suggested codes. This is rather problematic and other non-statistical methods will be needed to make up for this.

7 Future work

In the light of these results, the next step will be to conduct an experiment on categories as classes using stemmed extracted terms as features. This should improve further the 81.1% recall from the results of experiment 4.

These experiments were conducted in order to select the right features to be used as attributes for our machine learning module. We chose Naïve Bayes as a baseline method. However, other methods have been tested in previous works (Larkey and Croft, 1995) and have proved to give good results as well, such as k-nearest neighbors or Support Vector Machines.

We saw, at the end of section 6, that symbolic methods need to be developed in order to assist machine learning methods. Machine learning techniques have their limitations: they cannot assign codes for which they did not have enough data, and they cannot face the change to a new nomenclature. Therefore, in the near future we will have to develop a symbolic module comprising a series of linguistic rules in order to do the matching on all codes. A prototype based on the matching of morphemes has already been developed but will need to be experimented further.

The results of the experiments we conducted on a machine learning method were promising. Now, combining these two different approaches is the next challenging task in our project.

Acknowledgements

The CAPADIS project was funded by the government of the Brussels-Capital Region (Institute for the encouragement of Scientific Research and Innovation of Brussels) within the framework of the "Prospective research for Brussels 2006" program. I would also like to thank Benoît Debande, Claire Beguin, Sabine Regout and the Saint-Luc hospital team of coders.

References

Ananiadou S., McNaught J.: Introduction to Text Mining in Biology. In Ananiadou S., McNaught J. (eds.) Text Mining for Biology and Biomedicine, pp 1--12, Artech House Books (2006).

Aronson A. R.: MetaMap: Mapping Text to the UMLS Metathesaurus (2006).

Ceusters W., Michel C., Penson D., Mauclet E.: Semi-automated encoding of diagnoses and medical procedures combining ICD-9-CM with computational-linguistic tools. Ann Med Milit Belg;8(2):53—58 (1994).

Deville G., Herbigniaux E., Mousel P., Thienpont G., Wéry M.: ANTHEM: Advanced Natural Language Interface for Multilingual Text Generation in Healthcare (1996).

Farkas R., Szarvas G. Automatic construction of rule-based ICD-9-CM coding systems', BMC Bioinformatics, 9 (2008).

Friedman C., Shagina L., Lussier Y.A., Hripcsak G.: Automated encoding of clinical documents based on natural language processing. J Am Med Inform Assoc. 2004 Sep-Oct;11(5):392--402. Epub 2004 Jun 7 (2004).

Larkey L. S, Croft W. B. Automatic assignment of icd9 codes to discharge summaries. Technical report, University of Massachusetts at Amherst, Amherst, MA (1995).

Medori J. From Free Text to ICD: Development of a Coding Help, In: Proceedings of Louhi 08, Turku, 3-4 sept 2008 (2008).

Pakhomov S. V. S., Buntrock J. D., Chute C. G.: Automating the Assignment of Diagnosis Codes to Patient Encounters Using Example-based and Machine Learning Techniques (2006).

Paumier S. De la reconnaissance de formes linguistiques à l'analyse syntaxique. PhD thesis. Université de Marne-la-Vallée (2003).

Practice Management Information Corporation. ICD-9-CM Hospital Edition, International Classification of Diseases 9th Revision, Clinical Modification (Color-Coded, Volumes 1-3, Thumb-Indexed) (2005).

Pereira S., Névéol A., Massari P., Joubert M., Darmoni S.J. : Construction of a semi-automated ICD-10 coding help system to optimize medical and economic coding. Proc. MIE. (2006).

Pestian J. P., Brew C., Matykiewicz P.M., Hovermale D.J., Johnson N., Cohen K.B., Duch W.: A shared task involving multi-label classification of clinical free text. Proceedings of ACL BioNLP; 2007 Jun; Prague (2007).

Sager N., Lyman M., Nhán N., Tick L.: Medical language processing: Applications to patient data representation and automatic encoding. Methods of Information in Medicine, (34):140 -- 146 (1995).

Witten I.H., Frank E. Data Mining: Practical Machine Learning Tools and Techniques. San Francisco: Morgan Kaufmann Publishers. 2nd edition. 560 pp. ISBN 0-12-088407-0 (2005).

Zweigenbaum P. and Consortium MENELAS: MENELAS: coding and information retrieval from natural language patient discharge summaries. In Laires M. F., Ladeira M. J., Christensen J. P., (eds.), Advances in Health Telematics, pages 82-89. IOS Press, Amsterdam, 1995. MENELAS Final Edited Progress Report (1995).

Extracting Formulaic and Free Text Clinical Research Articles Metadata using Conditional Random Fields

Sein Lin[1] **Jun-Ping Ng**[1] **Shreyasee Pradhan**[2]
Jatin Shah[2] **Ricardo Pietrobon**[2] **Min-Yen Kan**[1]

[1]Department of Computer Science, National University of Singapore
justin@seinlin.com, {junping,kanmy}@comp.nus.edu.sg

[2]Duke-NUS Graduate Medical School Singapore
{shreyasee.pradhan,jashstar}@gmail.com, rpietro@duke.edu

Abstract

We explore the use of conditional random fields (CRFs) to automatically extract important metadata from clinical research articles. These metadata fields include formulaic metadata about the authors, extracted from the title page, as well as free text fields concerning the study's critical parameters, such as longitudinal variables and medical intervention methods, extracted from the body text of the article. Extracting such information can help both readers conduct deep semantic search of articles and policy makers and sociologists track macro level trends in research. Preliminary results show an acceptable level of performance for formulaic metadata and a high precision for those found in the free text.

1 Introduction

The increasing number of clinical research articles published each year is a double-edged sword. As of 2009, PubMed indexed over 19 million citations, over which 700,000 were added over the previous year[1]. While the research results further our knowledge and competency in the field, the volume of information poses a challenge to researchers who need to stay up to speed. Even within a single clinical research area, there can be hundreds of new clinical research results per year. Policy makers, who need to decide which clinical research proposals to fund and fast-track, and which proposals could tag onto existing research and cost share, have equally daunting information synthesis issues that have both monetary and public health implications (Johnston et al., 2006).

[1]http://www.nlm.nih.gov/bsd/bsd_key.html

Systematic reviews – secondary publications that compile evidence and best practices from primary research results – partially address these concerns, but can take years before their final publication, due to liability and administrative overheads. In many fast-paced fields of clinical practice, such guidelines can be outdated by the time of publication. Researchers and policy makers alike still need effective tools to help them search, digest and organize their knowledge of the primary literature.

One avenue that researchers have turned to is the use of automated information extraction (IE). We distinguish between two distinct uses of Information Extraction: 1) extracting regular, formulaic fields (*e.g.*, author names, their institutional affiliation and email addresses), and 2) extracting free text descriptions of key study parameters (*e.g.*, longitudinal variables, observation time periods, databases utilized).

Extracting such formulaic fields helps policy makers determine returns on health-care investments (Kwan et al., 2007), as well as researchers in large scale sociological studies understand macroscopic trends in clinical research authorship and topic shifts over time (Cappell and Davis, 2008; Lin et al., 2008). But due to the wide variety of publication venues for clinical research, even performing the seemingly simple task of author name extraction turns out to be difficult, and published studies thus far have relied on manual analysis and extraction.

Proposals to extract values of key study parameters may have more profound effects. Deeper characterization of research artifacts will enable more semantically-oriented searches of the clinical literature. Further programmatic access allows and encourages data sharing of raw clinical trial results, databases and cohorts (Piwowar and Chap-

Proceedings of the NAACL HLT 2010 Second Louhi Workshop on Text and Data Mining of Health Documents, pages 90–95,
Los Angeles, California, June 2010. ©2010 Association for Computational Linguistics

man, 2008) that may result in cost sharing across on-going studies, saving funds for other deserving clinical trials.

On one hand, the medical community has been proactive in using natural language processing (NLP) and information extraction technology in analyzing their own literature. Many approaches to metadata extraction have used regular expressions or baseline supervised machine learning classifiers. However, these techniques are not considered state-of-the-art.

On the other hand, much of the work from the NLP community applied to biomedical research has been on in-depth relationship extraction, such as the identification of gene pathways and protein-protein interaction (PPI). While certainly difficult and worthwhile problems to solve, there is room for contribution even at the basic IE level, to retrieve both regular and free form metadata fields.

We address this need in this paper. We apply a linear-chain Conditional Random Field (CRF) (Lafferty et al., 2001) as our methodology for extracting metadata fields. CRFs are a sequence labeling model that has shown good performance over a large number of information extraction tasks. We conduct experiments using basic token features to assess their efficacy for metadata extraction. While preliminary, our results indicate that CRFs are suitable for identifying formulaic metadata, but may need additional deeper, natural language processing features to identify free text fields.

2 Related Work

Many researchers have recognized the utility of the application of IE on biomedical text. These works have focused mainly on the application of well-known machine learning algorithms to tag important biomedical entities such as genes and proteins within biomedical articles. (Tanabe and Wilbur, 2002) uses a Naïve Bayes classifier, while (Zhou et al., 2004) uses use a Hidden Markov Model (HMM).

The Conditional Random Field (CRF) learning model combines the strengths of two well known methods: the Hidden Markov Model (HMM), a sequence labeling methodology, and the Maximum Entropy Model, a classification methodology. It models the probability of a class label y for a given token, directly from the observable stream of tokens x in direct (discriminative) manner, rather than as a by-product as in generative methods such as in HMMs. A CRF can model arbitrary dependencies between observation and class variables, but most commonly, a simple linear chain sequence is used (which connects adjacent class variables to each other and to their corresponding observation variable), making them topologically similar to HMMs.

Since their inception in 2001, (linear chain) CRFs have been applied extensively to many areas, including the biomedical field. CRFs have been used for the processing and extraction of important medical entities and their relationships among each other. (He and Kayaalp, 2008) reports on the suitability of CRFs to find biological entities, combining basic orthographic token features with features derived from semantic lexicons such as UMLS, ABGene, SemRep and MetaMap. In a related vein, CRFs have been applied to gene map and relationship identification as well (Bundschus et al., 2008; Talreja et al., 2004).

In a different domain, digital library practitioners have also studied how to extract formulaic metadata to enable more comprehensive article indexing. To extract author and title information, systems have used both the Support Vector Machine (SVM) (Han et al., 2003) and CRFs (Peng and McCallum, 2004; Councill et al., 2008). These works have been applied largely to the computer science community and have not yet been extensively tested on biomedical and clinical research articles.

Our work differs from the above by making use of CRFs to extract fields in clinical text. Similar lexical-based features are employed, however in addition to regular author metadata, we also attempt to extract domain-specific fields from the body text of the article.

3 Method

External to the scope of the research presented here, our wider project goal focuses on constructing a knowledge base of clinical researchers, databases, instruments and expertise in the Asia-Pacific region.

Dataset. In this pilot study, we created a gold standard dataset consisting of freely-available articles available from PubMedCentral. These articles focused on health services research in the Asia-Pacific region. In particular, we selected open-access full-text literature documenting oncological and cardio-vascular studies in the region, over a three year period from 2005 to 2008.

By constructing an appropriate staged query with PubMed, we obtained an initial listing of 260 articles. From an initial analysis, we determined that a significant portion (~1/3) of the retrieved full-text were not primary research, but reviews, case studies, editorials or descriptions. After eliminating these, the remaining 185 articles were earmarked to be manually tagged by clinicians affiliated with the project. Since the resulting corpus compiles articles across different journals and other publication venues, their presentation of even the formulaic author metadata varied.

The clinicians were given rich text (RTF) versions of the original HTML documents retrieved from PubMed. They identified and extracted only the sections of the articles that had pertinent data classes to tag. This process excluded most introductory, discussion and result sections, preserving the sections that described the study and results at a high level (*e.g., Demographics* and *Methods*).

After an initial training session, each clinician used a word processor to manually insert opening and closing XML tags for the tagset for a particular subsection of the 185-article corpus. Due to the high cost of clinician time, we chose to emphasize coverage, rather than have the clinicians multiply annotate the same articles. As a result, we could not calculate annotation agreement, but feel that the repeatability of the annotation was addressed by the initial training. At the time of writing, 93 articles have been completely tagged and sectioned, with the remainder in progress. The average length of the documents is about 1300 words. Once the dataset has been completed, we plan to release the annotated data offsets to the public, to encourage comparative evaluation.

The clinicians annotated the following *Formulaic Author Metadata* (3 classes):

- **Author (Au):** The names of the authors of the study;
- **E-mail (Em):** The email addresses of the corresponding authors of the study;
- **Institution (In):** The names of the institutions that the authors are from.

Such metadata can be used to build an author citation network for macro trend analysis. Note that this data is obtained from the article's title page itself, and not from any references to source articles, which have been the target of previous studies on CRF-based information extraction (Peng and McCallum, 2004; Councill et al., 2008). The clinicians also annotated the following *Key Study Parameters* (10 classes):

- **Age Group (Ag):** The age range of the subjects of the study (*e.g., 45 and 80 years, 21-79 years*);
- **Data Analysis Name (Da):** The name of the method or software used in the analysis of data collected for the study (*e.g., proportional hazards survival models, SAS package*);
- **Data Collection Method (Dc):** The data collection methods for the study (*e.g., medical records, review of medical records and linkage to computerized discharge abstracts*);
- **Database Name (Dn):** The name of any biomedical databases used or mentioned in the study (*e.g., Queensland Cancer Registry, National Death Index, population-based registry*);
- **Data Type (Dt):** The type of data involved in the study (*e.g., Cohort study, retrospectively*);
- **Geographical Area (Ga):** The names of the geographical area in which an experiment takes place or the subjects are from (*e.g., Pune, Switzerland*);
- **Intervention (Iv):** The name of medical intervention used in the study (*e.g., surgery, radiotherapy, chemotherapy, radio-frequency ablation*);
- **Longitudinal Variables (Lv):** Data collected over the observation period (*e.g., subjects*);
- **Number of Observations (No):** The number of cases or subjects observed in the study (*e.g., 158 Indigenous, 84 patients*);

- **Time Period (Tp):** The duration of an experiment or observation in the study (*e.g., 1997–2002, between January 1988 and June 2006*).

As can be seen from the examples, the tagging guidelines loosely define the criteria for tagging. For some classes, clinicians tagged entire noun phrases or clauses, and for others, only numeric values and modifiers were tagged. This variability arises from the difficulty in tagging these free text fields.

Features. The CRF model requires a set of binary features to serve as a representation of the text. A simple baseline is to use the presence/absence of particular tokens as features. The CRF software implementation we utilized is CRF++[2], which compiles the binary features automatically from a context window centered the current labeling problem instance.

We first preprocess an input article from its RTF representation and convert it into plain text. This is a lossy transformation that discards font information and corrupts mathematical symbols that could be helpful in the detection task. We take hyphenated token forms (*e.g., 2006-2007*) and convert them into individual tokens. The plain text is processed to note the specific locations of the XML tags for the learning process. The bulk of the words in each article were not tagged by clinicians, and for these words, we assigned a **Not Applicable (NA)** tag. We list the simple inventory of feature types that we use for classification.

- **Vocabulary:** Individual word tokens are stemmed with Porter's stemming algorithm (Porter, 1980) and down-cased to collapse orthographic variants. Each word token is then used as an individual feature. This feature alone was used to compile baseline performance as discussed later in evaluation.

- **Lexical:** Lists of keywords were compiled to lend additional weight for specific classes. In particular, we compiled lists of months, common names, cue words that signaled observations, institution names and data analyses methods. For example, a list of common given and surname names is useful for the **Au** field; while a list of months and their abbreviated forms help to identify **Tp**. Each list constitutes a different feature. As an example, in the case of human names, the names *Alice* and *Auburn* are on the list. If a word token corresponds to any of the words in the list, the corresponding feature is turned on (*e.g.*, isApersonName).

- **Position:** If a word token is within the first 15 lines of an article, this feature is turned on. This specifically caters to limit the scope of the formulaic author metadata fields, to match them only at the beginning of the article.

- **Email:** We create a specific feature for email addresses that is turned on when a particular word token is matched by a handwritten regular expression.

- **Numeric:** For some free text classes, such as **Ag**, **No** and **Tp**, the tagged text often contains numeric data. This can be present in both numeric and word form (*e.g., 23* versus. *twenty-three*). We turn this feature on for a token solely containing digits or numeric word forms.

- **Orthographic:** Orthographic features, such as the capitalization of a word token are useful to help identify proper nouns and names. If there are capital letters within a word token, this feature is turned on.

4 Evaluation

To ascertain the efficacy of our proposed solution, three-fold cross validation (CV) was first performed on a dataset comprising the 93 articles which have been completely annotated.

Baseline. For the purpose of comparison, we created a baseline system that utilizes the same CRF++ toolkit but uses only the vocabulary feature type with a five-word window (two previous tokens, the target token to be classified, and two subsequent tokens). The performance of this baseline system is shown in Table 1, where the standard classification performance measures of precision, recall and F_1 are given. *Count* measures the number of word tokens that are predicted as belonging to the stated field.

Discussion. We see that the overwhelming majority of tokens are not tagged (belonging to class **NA**).

[2] `http://crfpp.sourceforge.net`

The skewness of the dataset is not uncommon for IE tasks.

The baseline results show weak performance across the board. Clearly, significant feature engineering could help boost performance. Of particular surprise was the relatively weak performance on the formulaic metadata. From our manual analysis, it was clear that the wide range and variety of tokens present in names and institutions barred the system from achieving good performance on these classes. Comparative studies in citation and reference parsing usually peg classification performance of these classes at the 95% and above level.

Without suitable customization, detection of the key study parameters was also not possible. Only relatively common fields could be captured by the CRF, and when captured were more precise but lacked enough data to build a model with any acceptable level of recall.

Table 2 illustrates the improved results obtained by running CRF++ with all of the described features on the same dataset. The same five-word window size is used for the vocabulary feature. As seen, significant improvements over the baseline are obtained for all except four fields — **Da**, **Dc**, **Iv**, and **Lv**. These four fields were the classes with the most variability in annotation. For example, the data collection methodologies (**Dc**) and interventions (**Iv**) are often captured as long sentence fragments and hard to model with individual word cues.

The largest improvements occurred for the classes of age groups **Ag** and time periods **Tp**, both of which benefited from the addition of the numeric feature which boosted recognition performance.

Field	Prec.	Recall	F_1	Count
Formulaic Author Metadata				
Au	84.6	74.3	79.1	1818
Em	93.4	92.2	92.8	151
In	80.5	69.5	74.6	3906
Macro Avg.	86.2	78.7	82.3	
Key Study Parameters				
Ag	29.0	40.4	33.8	334
Da	61.0	39.0	47.6	708
Dc	8.3	3.2	4.6	48
Dn	35.9	15.1	21.2	92
Dt	52.8	26.8	35.5	36
Ga	7.3	4.5	5.6	41
Iv	4.6	1.4	2.1	22
Lv	15.4	20.0	17.4	13
No	14.4	5.8	8.3	125
Tp	73.6	55.8	63.5	261
Macro Avg.	30.2	21.2	24.0	
NA	97.1	98.5	97.8	119998

Table 1: Baseline aggregated results over 93 tagged articles under 3 fold cross validation.

Field	P.	Recall	F_1	Count
Formulaic Author Metadata				
Au	89.0	85.3	87.1	7312
Em	100.0	97.3	98.6	154
In	91.3	78.0	84.1	4515
Macro Avg.	93.4	86.6	89.9	
Key Study Parameters				
Ag	64.3	35.4	45.7	240
Da	79.3	37.2	50.6	2296
Dc	20.0	1.6	2.9	125
Dn	42.5	10.5	16.8	219
Dt	70.0	19.7	30.7	71
Ga	43.7	10.4	16.8	62
Iv	40.0	2.7	5.1	73
Lv	0.0	0.0	0.0	10
No	43.4	10.7	17.1	308
Tp	82.7	69.4	75.5	344
Macro Avg.	48.5	19.7	26.1	
NA	97.5	99.3	98.4	120430

Table 2: Aggregated results using the full feature set under 3 fold cross validation.

5 Future Work

The work presented here is ongoing, and based on our current results, we are planning to re-examine the quality of the annotations and refine our annotation guideline and scheme. We discovered cases where the CRF tagger correctly annotated key study parameters which the annotators had missed or mis-keyed. Drawing on lessons from the initial annotation exercise, a more comprehensive guideline is planned which will provide concise instructions with accompanying annotation examples.

We also plan to enrich the feature set. The current

set employed is still simplistic and serves as a developmental platform for furthering our feature engineering process. For example, the vocabulary, position and word lists features can be further modified to capture more fined-grained information.

Once we exhaust the development of basic features, our future work will attempt to harness deeper, semantic features, making use of part-of-speech tags, grammar parses, and named entity recognition for example. The incorporation of these features will likely be useful in improving the performance of the CRF learner. We also plan to use both clinical research and general medical ontologies (*e.g.*, UMLS) to gain additional insight on individual terms that have special domain-specific meanings.

6 Conclusion

We have developed a CRF-based information extraction system that targets two different types of metadata present in clinical articles. Our work in progress demonstrates that formulaic author metadata can be effectively extracted using the CRF methodology. By further performing feature engineering, we were able to extract key study parameters with a moderate level of success. Our post evaluation analysis indicates that more careful attention to annotation and feature engineering will be necessary to garner acceptable performance of such important clinical study parameters.

Acknowledgments

We like to express our gratitude to the reviewers whose insightful comments and pointers to additional relevant studies have helped improve the paper.

References

M. Bundschus, M. Dejori, M. Stetter, V. Tresp, and H.P. Kriegel. 2008. Extraction of semantic biomedical relations from text using conditional random fields. *BMC bioinformatics*, 9(1):207.

Mitchell S. Cappell and Michael Davis. 2008. A significant decline in the american domination of research in gastroenterology with increasing globalization from 1980 to 2005: An analysis of american authorship among 8,251 articles. *The American Journal of Gastroenterology*, 103:1065–1074.

Isaac G. Councill, C. Lee Giles, and Min-Yen Kan. 2008. ParsCit: An open-source CRF reference string parsing package. In *Proceedings of the Language Resources and Evaluation Conference (LREC 08)*, Marrakesh, Morrocco.

Hui Han, C. Giles, E. Manavoglu, H. Zha, Z. Zhang, and Ed Fox. 2003. Automatic document meta-data extraction using support vector machines. In *Proceedings of Joint Conference on Digital Libraries*.

Ying He and Mehmet Kayaalp. 2008. Biological entity recognition with conditional random fields. In *Proceedings of the Annual Symposium of the American Medical Informatics Association (AMIA)*, pages 293–297.

S.C. Johnston, J.D. Rootenberg, S Katrak, Wade S. Smith, and Jacob S Elkins. 2006. Effect of a us national institutes of health programme of clinical trials on public health and costs. *Lancet*, 367:13191327.

Patrick Kwan, Janice Johnston, Anne Fung, Doris SY Chong, Richard Collins, and Su Lo. 2007. A systematic evaluation of payback of publicly funded health and health services research in hong kong. *BMC Health Services Research*, 7(1):121.

John Lafferty, Andrew McCallum, and Fernando Pereira. 2001. Conditional Random Fields: Probabilistic Models for Segmenting and Labeling Sequence Data. In *Proceedings of the International Conference on Machine Learning*, pages 282–289.

JM Lin, JW Bohland, P Andrews, Burns GA, CB Allen, and PP Mitra. 2008. An analysis of the abstracts presented at the annual meetings of the society for neuroscience from 2001 to 2006. *PLoS ONE*, 3(e2052).

F. Peng and A. McCallum. 2004. Accurate information extraction from research papers using conditional random fields. In *Proceedings of Human Language Technology Conference and North American Chapter of the Association for Computational Linguistics (HLT-NAACL)*, pages 329–336.

Heather A. Piwowar and Wendy W. Chapman. 2008. Identifying data sharing in biomedical literature. In *Proceedings of the Annual Symposium of the American Medical Informatics Association (AMIA)*.

M.F. Porter. 1980. An Algorithm For Suffix Stripping. 14(3):130–137.

R. Talreja, A. Schein, S. Winters, and L. Ungar. 2004. GeneTaggerCRF: An entity tagger for recognizing gene names in text. Technical report, Univ. of Pennsylvania.

L. Tanabe and W.J. Wilbur. 2002. Tagging gene and protein names in biomedical text. *Bioinformatics*, 18(8):1124.

G. Zhou, J. Zhang, J. Su, D. Shen, and C. Tan. 2004. Recognizing names in biomedical texts: a machine learning approach. *Bioinformatics*, 20(7):1178–1190.

Author Index